'System change not climate change has been an increasingly strong rallying call within climate action movements. But, what does it really mean? In this second edition, Hans Baer details the massive implications of climate change, ten years on — in the climate emergency that we are now all experiencing.'

Anitra Nelson, *Small is Necessary* (2018), *Exploring Degrowth: A Critical Guide* (2020), *Beyond Money a Postcapitalist Strategy* (2022)

'An engaging and persuasive vision of Australia transformed by eco-socialism. Opening chapters explain the impacts of climate change on Australia and the economics and politics behind Australia's use of fossil fuels. The middle chapters show how mainstream politics has failed to deal with the crisis, as well as giving a detailed analysis of the environmentalist and leftist response. In a pause for reflection, Baer presents an auto-ethnography of activism on climate change, a great description of how these movements operate at the grass roots. Rod Quantock, a left comedian in Australia, is interviewed and explains the likelihood of collapse. This provides a platform to launch Baer's more optimistic vision of eco-socialism. The detail of these suggestions is impressive and will be much appreciated by all those concerned about our environmental crisis.'

Terry Leahy, *The University of Newcastle, Australia*

CLIMATE CHANGE AND CAPITALISM IN AUSTRALIA

Recognizing that climate politics has been an increasingly contentious and heated topic in Australia over the past two decades, this book examines Australian capitalism as a driver of climate change and the nexus between the corporations and Coalition and Australian Labor parties.

As a highly developed country, Australia is punching above its weight in terms of contributing to greenhouse gas emissions despite rising temperatures, droughts, water shortages and raging bushfires, storm surges and flooding, and the bleaching of the Great Barrier Reef. Drawing upon both archival and ethnographic research, Hans A. Baer examines Australian climate politics at the margins, namely the Greens, the labor union, the environmental NGOs, and the grass-roots climate movement. Adopting a climate justice perspective which calls for "system change, not climate change" as opposed to the conventional approach of seeking to mitigate emissions through market mechanisms and techno-fixes, particularly renewable energy sources, this book posits system-challenging transitional steps to shift Australia toward an eco-socialist vision in keeping with a burgeoning global socio-ecological revolution.

Accessibly written and including an interview with renowned comedian and climate activist Rod Quantock OAM, this book is essential reading for academics, students, and general readers interested in climate change and climate activism.

Hans A. Baer is an anthropologist based at the University of Melbourne. He works extensively on a diversity of research topics, including Mormonism, African American religion, socio-political life in East Germany, critical health anthropology, medical pluralism in the US, UK, and Australia, the critical anthropology of climate change, and eco-socialism.

CLIMATE CHANGE AND CAPITALISM IN AUSTRALIA

An Eco-Socialist Vision for the Future

Hans A. Baer

Routledge
Taylor & Francis Group

LONDON AND NEW YORK

First published 2022
by Routledge
2 Park Square, Milton Park, Abingdon, Oxon OX14 4RN

and by Routledge
605 Third Avenue, New York, NY 10158

Routledge is an imprint of the Taylor & Francis Group, an informa business

© 2022 Hans A. Baer

British Library Cataloguing-in-Publication Data
A catalogue record for this book is available from the British Library

Library of Congress Cataloging-in-Publication Data
Names: Baer, Hans A., 1944– author.
Title: Climate change and capitalism in Australia : an eco-socialist vision
    for the future / Hans A. Baer.
Description: Abingdon, Oxon ; New York, NY : Routledge, 2022. |
    Includes bibliographical references and index.
Subjects: LCSH: Climatic changes—Political aspects—Australia. | Climatic
    changes—Economic aspects—Australia. | Capitalism—Australia. |
    Environmental policy—Australia. | Environmentalism—Australia. |
    Environmental justice—Australia.
Classification: LCC QC903.2.A8 B34 2022 (print) | LCC QC903.2.A8
    (ebook) | DDC 363.738/7460994—dc23
LC record available at https://lccn.loc.gov/2021017063
LC ebook record available at https://lccn.loc.gov/2021017064

ISBN: 978-1-032-06489-5 (hbk)
ISBN: 978-1-032-06488-8 (pbk)
ISBN: 978-1-003-20253-0 (ebk)

DOI: 10.4324/9781003202530

Typeset in Bembo
by Apex CoVantage, LLC

CONTENTS

TABLES

FIGURE

PREFACE

As a health and ecological anthropologist, I started working on climate change issues starting around 2005 when Hurricane Katrina hit New Orleans, not far from where I was at the time situated as an academic at the University of Arkansas at Little Rock. Shortly after immigrating to Australia in January 2006 to start up a continuing academic position at the University of Melbourne, I have been working on climate-change-related issues full steam and have published, sometimes alone and sometimes with others, numerous articles, book chapters, and books that touch upon climate change and sustainability issues in one way or the other. Since 2008, at some level I have acted as a climate justice activist, seeking to engage in *praxis* or the merger of theory and social action. I have attended numerous climate rallies and climate action conferences in the larger society and also climate-change-related seminars, lectures, and panel discussions on climate-change-related issues, particularly at the University of Melbourne but also at other Australian universities. People in these settings almost uniformly accept the findings of climate science that anthropogenic climate change is already having devastating impacts on human populations and non-human species around the world, including Australia. They generally are not climate deniers or skeptics who one might find in the more conservative corridors of corporate and political life or the larger society. Nevertheless, while there may be a recognition that specific practices or activities, such as fossil fuel production and utilization, consumption, or meat eating, contribute to greenhouse gas emissions and thereby climate change, I have found that there is often a reluctance to situate these drivers under a larger driver, namely capitalism, a world system that has diffused to virtually every inhabitable corner of the world. I am not entirely sure why this is the case, but I suspect that it often is that people generally are deeply embedded in the system, and coming to terms with this might require drastic social and behavioral changes.

Within the Australian context, this book is written in the spirit of the adage "system change, not climate change" or more precisely "eco-socialism or catastrophic climate change." I suspect that this premise will not sit well with many parties who believe that a program of rapid decarbonization that shifts the world system, including Australia, from fossil fuels to renewable energy sources, with perhaps carbon pricing of some sort, will solve the climate crisis. However, just as capitalism has operated on energy sources, particularly fossil fuels, capitalism could operate on renewable energy sources without addressing its other impacts on the ecosystem and the gross social disparities that accompany it. Thus, at the end of the day, solving the climate crisis and related crises requires a radical socio-ecological revolution that transcends capitalism and replaces it with an alternative system based upon social justice, deep democracy, environmental sustainability, and a safe climate, both worldwide and in specific countries, such as Australia.

First of all, I would like to acknowledge the Anthropology and Development Studies Program at the University of Melbourne, which has provided me with a space to think about climate change-related topics, both globally and more specifically in Australia, since my arrival there in January 2006 as a full-time academic until January 2014, when I retired from full-time employment and continued on as a honorary fellow to the present time, still very much engaged as a semi-retired academic, engaged in some teaching, supervision of postgraduate students, and particularly research. In terms of researching climate-change-related topics, I would like to acknowledge Merrill Singer, with whom I co-authored two books, namely *Global Warming and the Political Ecology of Health* (Baer and Singer 2009) and *The Anthropology of Climate Change* (Baer and Singer 2014, 2018), and Verity Burgmann, with whom I co-authored *Climate Politics and the Climate Movement in Australia* (Burgmann and Baer 2012). In addition, numerous scholars and activists too many to list have sharpened my thinking on the impact of climate change on Australia, the drivers of climate change in Australia, and eco-socialism and other radical perspectives. They include Samuel Alexander, Sue Bolton, Jim Falk, Arnaud Gallois, James Goodman, Pip Hinman, Suzi Hutchings, Terry Leahy, Debbi Long, Jonathan Marshall, Anitra Nelson, Susan Price, Jack Roberts, Kim Sawyer, Ariel Salleh, David Spratt, Michael Stevenson, and Ted Trainer. I also acknowledge the space that my colleagues in the Melbourne Sustainable Society Institute Honorary Fellows Group have provided to air some of the topics covered in this book as well as others related to climate change and sustainability issues in our quarterly meetings at the University of Melbourne. I also acknowledge various groups with whom I have been involved in one capacity or other in my observations of the Australian climate movement, particularly the Socialist Alliance, Solidarity, the Australian Greens, the Climate Emergency Network, Friends of the Earth, Climate Action Moreland, the Climate Reality Project, and Psychology for a Safe Climate. Last but not least, I acknowledge permission by the Society of Applied Anthropology to allow me to republish the material in my article titled "Efforts to update the climate emergency framework: From Australia to the world and back to Australia," which appeared in the Winter 2021 issue of *Practicing Anthropology*, which appears in Chapter 5 of this book.

INTRODUCTION

Climate scientists and other observers often refer to various regions, such as the Arctic, low-lying islands in the South Pacific, the Andes, and Bangladesh, inhabited by indigenous and peasant peoples as the canaries in the coalmine, signaling the adverse impacts of anthropogenic climate change. It is often said that those people who have contributed the least to greenhouse gas emissions are the ones suffering the most from climate change, a more than accurate observation.

Multinational corporations based primarily in advanced capitalist societies and state and joint-venture companies in post-revolutionary societies, particularly China, have created a global factory and a new global ecosystem which includes industrial and motor vehicle pollution, toxic and radioactive wastes, deforestation, desertification, and climate change (Baer 2012; Smith 2020). As a result of its ever-expanding production, global capitalism has had adverse impacts on large sections of humanity and a fragile ecosystem. The Industrial Revolution relied first on coal, then petroleum, and finally natural gas, including coal seam gas, the burning of which all produces carbon dioxide emissions, which contribute to anthropogenic climate change.

A growing number of climate scientists argue that if humanity does not start to drastically reduce greenhouse gas emissions soon, there is a strong possibility that the average global temperature will be 4°C or higher than that at the time of the Industrial Revolution, a point at which the average global temperature already stands 1.1°C higher than that in 1900 (Lynas 2020). Ironically, the COVID-19 pandemic has resulted in a modest reduction of greenhouse gas emissions due to governmental lockdowns around the world temporarily shutting down or curtailing various industries and businesses, including air transportation, not so much for cargo but for passengers. However, the powers that be in both the corporate and state sectors are biting at the bit to return to business-as-usual or, in the words of Australian Prime Minister Scott Morrison, "snap-back." In large part in seeking

DOI: 10.4324/9781003202530-1

to address the COVID-19 pandemic, climate change has assumed a backseat for many policymakers, despite ongoing catastrophic climate events in various parts of the world.

Conversely, there is one country, a developed one at that, namely Australia, that is punching above its weight in terms of contributing to greenhouse gas emissions, not only from domestic consumption but due to massive export of coal, that constitutes a canary in the coalmine, signaling the adverse impacts of climate change. Over the past decade or so, Australia, as the driest settled continent, has been experiencing risking temperatures, droughts accompanied by water shortages and raging bushfires, storm surges and flooding, and the bleaching of the Great Barrier Reef.

The year 2020 tied with 2016 as the hottest year on record for both the globe and 2019 constituted the hottest year on record in Australia. Globally 20 of the 21 hottest years on record have occurred during the 21st century. Climate change and its impacts are becoming more and more dire and range from heat waves to droughts, wildfires or bushfires in North America, the Arctic, cyclones and hurricanes in many places around the world, the retreat of the Arctic icecap inducing polar vortexes resulting in freezing temperatures in most of the United States and Greece in early February 2021, a reduction in the size of ice caps in Antarctica and Greenland, glacial retreat in the Himalayas, Alps, Andes, and Rocky Mountains, rising sea levels, the acidification of the oceans, the slowing down of the North Atlantic Conveyor Belt that contributes to a warming effect on northwestern Europe, and so on.

In Australia, four degrees of global warming is predicted to dramatically alter rainfall and temperature patterns and distribution, leading to increased incidence and intensity of extreme weather events like droughts, heatwaves, floods, and storms. In 2019–2020 Australia experienced the most extensive bushfires on record, in which 21 percent of Australia's temperate broadleaf and mixed forests burnt, impacting either directly or indirectly nearly 80 percent of Australians, resulting not only in the devastation of numerous rural communities in southeastern Australia but also Sydney, Canberra, and Melbourne being blanketed by dangerous levels of smoke (Climate Council 2020). Furthermore, an estimated 1 billion animals were killed by the bushfires, 800,000 of them in New South Wales. Needless to say, various climatic events, whether they be rising temperatures, droughts, bushfires, cyclones, or flooding, will over the course of the 21st century have their most adverse impact on Indigenous Australians in capital cities, regional cities and towns, and remote areas.

Australians often believe that they reside in a "lucky country" which enjoys one of the highest standards of living in the world, although this characterization does not apply for most Indigenous people and a growing number of non-Indigenous Australians as the country experiences a growing amount of social inequality related to corporate globalization and neoliberalism. While Australians have experienced droughts and bushfires often in the past, the intensity of these events and other climate-related events, such as cyclones, storm surges, and flooding, has become much more pronounced.

Anthropogenic climate change and other ecological crises compel us to seriously confront not only its impact on Australia but the rest of humanity and non-human fauna and flora. Ultimately, it compels us to transcend the capitalist world system and begin a revolutionary eco-socialist shift to an alternative world system based upon social justice and parity, deep democracy, environmental sustainability, and a safe climate. While I have attempted to address this issue on a global level, in this book I turn my attention to my adopted country, namely Australia, in seeking to grapple with the possibilities of applying an eco-socialist framework to that country (Baer 2018).

Australia to those who don't live here might look like a place that's great despite itself. Its vast lands and resources have spared most of its inhabitants, other than Indigenous Australians, the upheaval, civil wars, abject poverty, and authoritarian controls endured by many in most former European colonies, including the United States. Indeed, many Americans superficially picture Australia as the home of Crocodile Dundee, kangaroos, cute koalas, beaches, the 2000 Sydney Olympics, and the outback populated by Indigenous people and rugged sheep farmers and cattlemen. Some even confuse it with Austria. Small wonder when American mainstream media generally accord Australian events fleeting coverage, if any. Also, despite its great physical size, Australia has been a relatively small player on the global geopolitical scene, contrary to what many Australian politicians and other Australians would like to believe. Yet, these perceptual divides aside, Australia constitutes an interesting site for contemplating what an anti-capitalist alternative might look in a country socio-politically situated between the United States, northern Europe, and Canada.

Consequential mistakes in valuing labor and in using and distributing resources made in recent decades are already jeopardizing the future, calling for dramatic changes in how we treat both people and the environment. How that change might look and occur is explored here in the hopes of kindling discussion and action. In our work as critical health anthropologists, I along with Merrill Singer and Ida Susser developed the notion of *democratic eco-socialism* as a strategy for developing a healthy world. For us, democratic eco-socialism comprises the following components:

- an economy oriented to meeting basic social needs for everyone, namely, adequate food, clothing, shelter, health care, and dignified work;
- a high degree of social equality;
- public ownership of the means of production;
- environmental sustainability and a safe climate (Baer, Singer, and Susser 1997, 2003, 2013).

We conflated the terms *democratic socialism* and *eco-socialism* in coining the term *democratic eco-socialism*, mainly in that we wanted to highlight that authentic socialism by definition must be democratic and to set it apart from efforts during the 20th century to develop socialism in numerous countries, starting out with the Bolshevik Revolution in 1917. In our view, these countries constituted socialist-oriented or post-revolutionary societies but never fully achieved socialism for complex historical and social structural, internal and external factors. Aside of our own work, the term *democratic eco-socialism* has not entered general discourse, but world-renowned Naomi Klein (2019: 251) recently referred to it, noting that a "new form of democratic eco-socialism, with the humility to learn from Indigenous teachings about the duties to future generations and the interconnection of all life, appears to be humanity's best shot at collective survival." Despite this, I am opting for purposes of this book to simply use the term *eco-socialism* with the understanding that socialism by definition needs to be democratic and to act in solidarity with the growing number of eco-socialist or ecological Marxist scholars and activists around the world (Loewy 2015; Albritton 2019; Thornett 2019; Kovel and Quincey 2019; Foster and Clark 2020).

While the notion that eco-socialism may be implemented in Australia or any other developed society may seem starry-eyed, history tells that social changes can occur very rapidly once economic, political, and social factors have reached a tipping point. Fred Magdoff and John Bellamy Foster (2011: 132) suggest that such a time is nearing: "Everywhere radical, essentially anti-capitalist, strategies are emerging, based on other ethics and forms of organization, rather than the profit motive." As Frank Stilwell (2000: 134) observes, a "socialist alternative must embody inspirational ideals while also offering down-to-earth proposals for dealing with the problems of contemporary Australian capitalism."

Beyond that, we can't forget that Australia's renowned luck could run out. In his now classic *The Lucky Country* (1964), journalist Donald Horne wryly insinuated that Australia managed to be "lucky" despite having had mostly mediocre political and economic leaders. And the "lucky country" now faces numerous crises, especially the two discussed below: increasing social inequality and erosion of workers' rights; and an ecological crisis presently manifested by the impacts of anthropogenic climate change and ultimately a by-product of global capitalism's emphasis on profit-making, endless economic expansion, and the treadmill of production and consumption heavily reliant on fossil fuels (Baer 2012; Klein 2014). As one of the hottest and driest continents, Australia is extremely vulnerable to climate change. Ironically, as the world's largest coal exporter, it is exceptionally reliant on fossil fuels for domestic consumption and export earnings, and its greenhouse gas emissions per capita rank among the highest in the world.

While Australia has been described as the 51st state because it has been Americanized in so many ways, particularly since World War II, it retains its own distinct identity. That identity is colored by its parliamentary system with the Queen of England as its head of state. Also, compared to the United States, Australia has a stronger (if eroding) welfare system, greater emphasis on multiculturalism and greater attention to indigenous peoples, greater secularism as manifested by electing some self-proclaimed atheists as prime ministers, and more subdued patriotism. Although this take is undercut somewhat by the strengthening of the US-Australian military alliance in recent years, particularly since 9/11, Australian political scientist Dennis Altman is basically correct that:

> Australia is not, as the United States has been since World War II, a militarised society. We have a far smaller military establishment and far less employment depends upon military expenditures.
>
> *(Altman 2006: 20–21)*

Australia's primary political, economic, military, and cultural allegiances were for long to Great Britain. Conversely, like New Zealand and unlike the United States and Canada among British settler colonies, Australia is so far from its mother country that "tyranny of distance" has come to explain its position in the capitalist world system. Since World War II, Australia has oriented itself increasingly to the United States, even following it into wars in Korea, Southeast Asia, Afghanistan, and Iraq. And Australia is much more integrated into the Asia Pacific region than are the United States and Canada, with their strong North Atlantic regional orientation.

In my dual roles as a scholar pursuing a critical anthropology of climate change and a climate justice activist, I have often felt marginal in both the academic world and the larger Australian society, where I have resided as an immigrant since January 1, 2016, after having spent a one-year visiting stint at the Australian National University in 2014, in my advocacy of eco-socialism as an alternative to global capitalism. Yet, repeatedly, I have often been struck how often many of my fellow academics who are also interested in climate politics, most of whom view themselves as progressive people, have resigned themselves to the belief that climate change mitigation ultimately must be solved within the parameter of the "market" or capitalism, whether in the Australian context or in the global context. Simply adopting capitalist solutions to the many contradictions of the capitalist world system, including green capitalism, is a kind of fool's paradise that misdiagnoses both the extent and the ultimate sources of facing the eco-health of the planet and its inhabitants, both human and non-human.

Eco-socialism as a vision for the future

My own sense is that in terms of climate change, things will continue in the foreseeable future for most countries in the world, both developed (including Australia) and developing, will get worse before they get better, if they ever get better, given that humanity has embarked upon a trajectory to "fry the planet." Drawing upon climate scientific observations, Clive Hamilton, an Australian public intellectual, in *Requiem for a Planet*, observes:

> Without some unforeseeable stroke of luck, a warming of four degrees and more appears very likely. The best estimate is that we will reach that

level in the 2070s or 2080, although if things go badly is could be as soon as the 2060s.

(Hamilton 2010: 196)

While I certainly hope that the warming of the planet will not prove to be this bad and do not want to appear like a doomsday prophet, it is imperative that we come to recognize as soon as possible, the sooner the better and the later the worse, that the existing global economy or the capitalist world system is running its course. As grim as Hamilton (2010: 226) sees the future of humanity as it proceeds further and further in the 21st century, he urges us not to despair but to move forward, ending his book with the counsel that "only by acting, and acting ethically, can we redeem our humanity." While on this note I am in total agreement with Hamilton, he does not leave use with a vision for getting humanity, including Australia, out of the quagmire that climate change presents all of us on numerous fronts.

Bearing this thought in mind, in this book I radically "push the envelope" by proposing an eco-socialist vision of an alternative Australia. I examine system-challenging or radical steps for shifting Australia from looming crises toward a democratic eco-socialist system. Since socialism can't work if confined to a single country, radical transitional reforms should be viewed as loose guidelines for shifting Australia away from a capitalist liberal democratic system and toward an eco-socialist system. Shifting toward an eco-socialist framework will have to be part and parcel of a global social transformation toward greater social parity, greater democratic processes, and environmental sustainability – all including more comprehensive and radical climate regimes than today's. Even if such a global transformation seems far off now, as the global socioeconomic and ecological crises unfold, it is up to progressive people to envision radical alternatives to the capitalist world system, particularly in its neoliberal phase. We need to have worldwide, regional, national, and even local plans.

Ultimately, envisioning implementing an eco-socialist vision in Australia must be part and parcel of a global socio-ecological revolution. Michael Loewy, a long-time eco-socialist, observes:

> The ecological/socialist revolution requires a radical break with the whole capitalist paradigm of civilization, with its ecologically destructive forms of production and consumption, and its unsustainable way of life. In other words, the traditional Marxist concept of revolution is indispensable, but has to be deepened, radicalized, and broadened. It has to include not only a radical change in the relations of production (private property), but also in the structure of the forces of production in the sources of energy (e.g., solar instead of fossil), and in wastefully patterns of consumption. Revolution now means the establishment of a new civilization which is leading humanity to an unprecedented catastrophic global warming.

(Loewy 2020: 138)

John Bellamy Foster (2009), another renowned eco-socialist, has been writing and speaking for over a decade about the need for a global ecological revolution. He argues:

> Traditional working-class politics are thus co-evolving and combining with environmental struggles, and with the movements of people of colour, of women, and all those fighting basic, reproductive battles throughout society. Such an ecological and social struggle will be revolutionary to the extent that its force from those layers of society where people's lives are most precarious: Third-World workers, working-class women, oppressed people of color in the imperial core, indigenous populations, peasants/landless agricultural workers, and those fighting for fundamentally new relations of sexuality, gender, family, and community – as well as highly exploited and disposed workers everywhere.
>
> *(Foster 2020a: 192)*

It is imperative to construct an alternative to global capitalism as the ultimate climate change mitigation strategy, even though it will not be achieved anytime soon, if ever for that matter. However, as humanity enters an era of increasingly dangerous climate change accompanied by tumultuous environmental and social consequences, it will have to consider alternatives that hopefully will circumvent dystopian scenarios. Eco-socialism, which draws on other perspectives, including eco-feminism and eco-anarchism, offers a vehicle for creating a safe climate, but also a more socially just, democratic, and environmentally sustainable world system with manifestations in the many regions and nation-states across the globe, including Australia and greater Australasia. While Foster (2020a: 192–193) anticipates that the ecological revolution, what I prefer to term a socio-ecological revolution, "can be expected to arise first in the Global South whose multitudes are the most exploited and oppression globally," he observes that these conditions are "being generalized across the entire world system as a result of spiralling economic and ecological rises, coupled with growing destruction and expropriation of the means of life themselves."

While at the present time or for the foreseeable future, the notion that eco-socialism may be eventually implemented in any society, developed or developing, or a number of societies may appear absurd, history tells us that social changes can occur very quickly once certain social structural and environmental conditions have reached a tipping point, a term that has also become popular in climate science. As humanity lurches ever forward into the 21st century, our survival as a species appears to be more and more precarious, particularly given that the impact of climate change in a multiplicity of ways looms on the horizon. Global, regional, national, and local temperature records are repeatedly being broken around the world.

Since late 2019, climate change and the broader socio-ecological crisis have been relegated to the sidelines due to the COVID-19 pandemic, both globally and in Australia. Malm (2020) in *Corona, Climate, Chronic Emergency* argues, however, that there is a complex interaction of global capitalism, climate change, and COVID-19. He observes that of the ten countries, namely the US, Italy, China, Spain, Germany, France, Iran, UK, Switzerland, and the Netherlands, with the most deaths from COVID-19 as of late March 2020, only Iran and Switzerland were not in the top listing of countries responsible for the most cumulative CO_2 emissions since 1751. Capitalist-driven deforestation and food production, particularly of wild animals, have resulted in zoonotic spill-over from animals to humans, in many cases starting with bats and intervening wild animals, such as possibly pangolins, sold in wet markets, particularly the one in Wuhan. EU countries source over half of their land-based consumption from other regions of the world, over 80 percent in the case of the Germany. At the high-end of capitalist developed countries, one finds Japan, which sources 92 percent of its land-based consumption from other regions, and at the low-end one finds the United States, which sources only 33 percent of its land-based consumption from other regions, largely because of its enormous land mass. Initially, the major epicenters of COVID-19 were generally rich cities, such as New York, London, and Hong Kong, but in due time megacities in the Global South, including in India, Brazil, Peru, and Mexico became epicenters of the pandemic, a scenario that is still playing itself out before our very eyes.

Malm argues that the three coronavirus epidemics might be the by-products of climate change. SARS appeared in the wake of a drought in Guangdong, MERS first appeared in rain-free Jedda, and SARS-COV-2 erupted amid the worst drought in Wuhan area in 40 years. Thus the combination of infectious disease epidemics and climate change suggests humanity has entered an era of chronic emergency. Malm (2020: 121) asserts that humanity faces two options: the first entailing flowing "deeper into catastrophe" and the second entailing transformation into "another form of socialism, one that recognises that time is up and another decade or even year of this status quo is intolerable." He distances himself from classical Marxists, who have viewed socialism as a social formation of unlimited abundance. Invoking Lenin's passion for wild nature, Malm (2020: 167) calls for, in a figurative sense, "ecological war communism," which would require

> learning to live without fossil fuels in no time, breaking the resistance of dominant classes, transforming the economy for the duration, refusing to give up even if all the worst-case scenarios come true, rising out of the ruins with the force and compromises required, organising the transitional period of restoration, staying with the dilemma.

He leaves it to others to figure out how humanity might transition from the existing capitalist world system that exploits not only people but also nature to ecological war communism. Whereas I have attempted to something along these lines for the world as a whole in my book *Democratic Eco-Socialism as a Real Utopia* (Baer 2018), in this book I attempt to envision how an eco-socialist vision might develop in one specific country, namely Australia, where I have resided on a permanent basis since 2006 and which has already been ravaged by the impact of climate change, some of which I have personally witnessed.

Chapter 1 (The Impact of Climate Change in Australia) recognizes that, as has been the case globally, the first two decades of the 21st century have been the warmest period on record in Australia. Heat events have become longer, more frequent, and more intense over much of Australia in recent decades. Indeed, 2019 proved to be the warmest year on record and Australia. In Australia, four degrees of warming is predicted to dramatically alter rainfall and temperature patterns and distribution, leading to increased incidence and intensity of extreme weather events like droughts, heatwaves, floods, and storms. These changes will adversely affect Australia's biodiversity, ocean, and agricultural systems, with potentially severe social and economic impacts, particularly in cities, where infrastructures, service provision, and patterns of settlement could face significant disruption. Furthermore, Australia will need to grapple with environmental problems within its broader region, including, for instance, food security and eco-refugee migration. Due to both climate change and non-climate-related stresses, the Great Barrier Reef has lost 50 percent of its coral cover of the past three decades. This chapter reviews four climatic events that have occurred in Australia in recent times, namely the Millennium Drought of 1997–2009; the period of 2010–2011, during which the Queensland and New South Wales coast experienced intense cyclones, storm surges, heavy rainfall, and flooding; the Angry Summer of 2012–2013; and the drought and intense bushfires that swept across states in late 2019 and early 2020. It also examines the projected impact of rising temperatures on Australia's capital cities over the period of 2015–2090 as well as the impact of climate change on Indigenous peoples, Australia's most marginal socio-economic group, particularly in remote areas.

In Chapter 2 (Australian Capitalism as a Driver of Climate Change), I examine various industries and social practices that contribute to greenhouse gas emissions and ultimately climate change in Australia. Within the context of the capitalist world system, Australia constitutes a developed capitalist country, according to some theorists a minor core country and other theorists an advanced semi-peripheral in a global system of unequal economic exchange. While it is situated in the southern hemisphere, it belongs to Global North along with the United States, Japan, Germany, the UK, France, and other advanced capitalist countries. From a world systems theory perspective, wealth and poverty are intricately interrelated, with core countries exploiting both semi-peripheral and peripheral countries and semi-peripheral countries being exploited by the core and in turn exploiting the periphery.

Despite its relatively small population of some 25.7 million inhabitants, Australia had made a disproportional contribution to climate change because both its ecological and greenhouse gas emissions are quite large. Australia's high level of greenhouse gas emissions is driven by its heavy dependence on fossil fuels for energy, particularly coal; aluminum smelting industry heavily dependent upon coal-generated electricity; heavy reliance on private automobiles and airplanes for both domestic and international travel; agricultural and forestry industries; a growing number of large office buildings; and a culture of consumption, which includes large dwelling units filled with numerous energy-dependent appliances and electronics products. While Australia's emissions are among the highest per capita in the world, national statistics are misleading in that they don't include emissions from the off-shore production of manufactured products imported to Australia from various developing countries, particularly China. Australia not only generates carbon dioxide emissions from its heavy reliance on coal, albeit diminishing due to the uptake of solar and wind power, but it outsources it, having been the first or second leading exporter of coal, particularly to Japan, South Korea, Taiwan, and China, and more recently the world's leading exporter of liquid natural gas.

Chapter 3 (Climate Politics at the Top: The Corporations and the Two Major Parties) examines the nexus between the corporate sector and the state in the Australian political economy. Globally, Australia is distinct in that it has been the only country that adopted and then abandoned some form of carbon pricing in the form of the Carbon Pricing Mechanism implemented by the Gillard ALP government with support of the Greens. In this chapter, I explore Australian climate politics at the top, particularly in the relationships between the corporate sector, particularly the mining and energy production sectors, and the two major political parties, namely the Coalition Party and the Australian Labor Party (ALP). Sometime ago Noam Chomsky said the United States has a "one party system," namely the Business Party with two factions, the Republicans and the Democrats. In a similar vein, one might argue that Australia also has a Business Party, with the Coalition and the Australian Labor Party being its factions. While prior to perhaps the 1980s, the ALP, with its strong alliance with the Australian labor movement, constituted an oppositional force to the capitalist class, with the full-fledged acceptance of neoliberalism by the ALP Hawke and Keating governments between 1983 and 1996, the ALP entered into an unholy alliance with sections of the Australian capitalist or corporate class.

In this chapter, I examine climate change political machinations at the top, namely between the corporations and the various Australian governments and their oppositions during the following periods: (1) the ALP governments and the Coalition opposition, 1983–1996; (2) the Coalition government and the ALP opposition, 1996–2007; (3) the ALP governments and the Coalition opposition, 2007–2013; and (4) the Coalition governments and ALP opposition, 2013–present. Climate change has adversely impacted the career of every Australian prime minister since John Howard, along with him during his later years in this role. Both Coalition and ALP governments at both the federal and state levels have consistently subsidized the exploitation of Australia's coal, natural gas, and coal seam gas supplies and supported the expansion of coal exports and liquid natural gas. The fossil fuel- and coal-based electricity generation industries have dominated almost every greenhouse-related consultative committee by the federal government and its agencies. This was also the case for the Coalition government of John Howard, the federal ALP governments under Prime Ministers Kevin Rudd and Julia Gillard between 2007 and 2013, and in subsequent Coalition governments with Tony Abbott, Malcolm Turnbull, and Scott Morrison at the helm. Over the years, the fossil fuels industry has systematically contributed to the major Australian political parties to curry favor to its economic interests.

I also examine the phenomenon of climate denialism as a link between the corporate sector and particularly the Coalition Party as it has played itself out in the activities of the Lavoisier Group and the Institute of Public Affairs. Conversely, as the evidence for climate change has become more apparent and acknowledged in the more progressive sectors of the mass media and on the part of even selected conservative politicians, Malcolm Turnbull being a case in point in the Australian context, in order to maintain a favorable public face, more and more corporations have adopted what might be termed a stance of *corporate environmentalism*, in some ways a more recent variant of the notion of *corporate social responsibility*.

In Chapter 4 (Climate Politics at the Margins: The Greens, Labor Unions, Environmental NGOs, and the Grass-Roots Climate Movement), I examine the policies and actions of the Greens as the largest minority party, the labor movement, the peak environmental NGOs, and the grassroots climate movement in Australian climate politics. Since the beginning of the 21st century, a climate movement, in part an outgrowth of an earlier environmental movement, emerged as the gravity of anthropogenic climate change has become recognized in many quarters of Australian society. Around the world, many environmental groups and Green parties have become involved in the climate movement as have other political actors, including grassroots climate action groups, socialist parties and groups, labor unions, religious bodies, and more progressive politicians. Given that corporations and most governments along with the United Nations have not been acting in a responsible and effective manner in terms of serious climate change mitigation, despite considerable rhetoric to the contrary, much of the collective effort of climate change action has been spurred by a still burgeoning international climate movement that is quite variable in terms of addressing social justice or equity issues and challenging global capitalism.

In this chapter, I examine the relationship between the Green Party, which, in contrast to the two major parties, has adopted a policy of leaving coal in the ground, and the Australian climate movement. I also present an overview of various layers in the Australian climate movement: (1) a mainstream layer consisting of the labor unions and various thinks, particularly the Climate Council; (2) environmental NGOs, particularly the Australian Conservation Foundation, Greenpeace, and Friends of the Earth; and (3) a grass-roots layer consisting of climate action summits, climate action groups, Rising Tide, socialist groups with a strong anti-capitalist and eco-socialist stance, the Lock the Gate Alliance, the Stop Adani campaign, and most recently the School Strike 4 Climate Action and Extinction Rebellion.

In Chapter 5 (Engaging with the Australian Climate Movement: An Autoethnography of the Climate Justice Activist), I present an autoethnography of my involvement as a climate justice activist in that movement. Autoethnography is a research method which relies upon a researcher's personal experience to describe and critique cultural beliefs and social structures and entails intense self-reflection or reflexivity. Since 2008 I have attended numerous climate rallies and climate action conferences, and worked in one capacity or other with various climate action groups, including Climate Action Moreland, the Climate Emergency Network in Melbourne, Climate Action Moreland, the Socialist Alliance, Psychology for a Safe Climate, and Al Gore's Climate Reality Project. I present accounts of my observations at selected climate rallies, climate action conferences and meetings, including the National Climate Emergency Summit in Melbourne in February 2020.

This chapter also presents a life history of Rod Quantock, a renowned left-wing Australian comedian with over 50 years of experience in stand-up comedy, cabaret, theatre, television and radio. Between April 7, 2018, and January 24, 2019, I conducted six in-depth, semi-structured interviews with Rod for purposes of a life history of an icon in both comedy and Australian social activism, particularly his climate activism over the past 15 years or so. My first five interviews focused on an assortment of topics, such as his upbringing, studies as an architecture student, involvement in political protests during the late 1960s, his career as a professional comedian, a stint as a columnist for *The Weekend Age*, and his evolution into a climate activist. In seeking to unravel Rod's *emic* or insider's perspective in anthropological parlance, I discuss his views on capitalism and the culture of consumption, politicians and political parties, climate change, and changing the system. I return to Rod's views on climate action in Chapter 6, where we engage in a dialogical conversation about our respective views on catastrophic climate change and climate action.

Chapter 6 (System-Challenging Transitional Steps to Shift Australia Toward Eco-Socialism) opens with a dialogic conversation with Rod Quantock based on our sixth interview revolving around his climate change dystopian views, bordering on catastrophism, and my own vision of democratic eco-socialism as a real utopian response to the socio-ecological crisis. In addition to having come of age more or less during the same era, namely the Sixties, he in Australia and I in the United States, Rod and I started to become radicalized during this decade and the subsequent decade. Whereas he shifted from studies in architecture to a career as a comedian, I shifted from work as an engineer in the aircraft industry to a career as an anthropologist. What Rod as a comedian and I as an anthropologist share is that we have both become climate change communicators, albeit he through the lens of a left-wing comedian and I through the lens of a critical anthropologist.

Following from my dialogic conversation with Rod, I briefly discuss various mainstream and radical perspectives on strategies for moving Australia to a more environmentally sustainable and climatically safe space. I explore system-challenging transitional steps needed to shift Australia toward an eco-socialist vision. In doing so, I discuss various eco-socialist visions, including socialist eco-feminism, as well as eco-anarchist and de-growth visions on how to transcend capitalism and its economic growth paradigm. I also seek to point out how in various ways particularly eco-socialism and eco-anarchism share with Indigenous local knowledge concerning human-nature relationships. While I do not intend to present a blueprint as such for moving Australia toward eco-socialism, I seek to present rough guidelines designed to provoke discussion and break the impasse encountered by many eco-socialists on how to go from A to B, namely the existing capitalist system to eco-socialism as a global phenomenon, with multiple national expressions, including in Australia.

In Chapter 7 (Australia's Role as a Climatically Vulnerable Country in the Larger Context of a Global Socio-Ecological Revolution) explores my proposal of an eco-socialist agenda for Australia as part of a larger global call for eco-socialism. Is my vision of an eco-socialist agenda for Australia wishful thinking? Perhaps the vision won't unfold by itself or soon. But if Australia, like other wealthy countries, keeps up its wholehearted embrace of the capitalist growth paradigm that's behind deepening social divides and a climate crisis worsening despite the deliberations at 26 UN conferences and despite emissions-trading schemes around the world, the alternative may be dystopia. Achieving an eco-socialist Australia would have to be part of a larger global effort in which a greatly enlarged climate justice movement works in alliance with other anti-systemic movements, such as the global justice, the labor, Indigenous and refugee rights, and women's rights movements. The Australian climate movement could more fully join in the struggle for social justice both globally and at home if it were to place greater emphasis on climate justice and less on ecological modernization, which has a role to play in climate change mitigation but will not counteract the capitalist treadmill of production and consumption and its drive for continual economic growth. Given the failure of existing international and national climate regimes to date in adequately mitigating climate change, this effort will have to be spurred from below.

1

THE IMPACT OF CLIMATE CHANGE IN AUSTRALIA

The capitalist world system has transformed humanity into a geological force, ushering in a new geological epoch: what Earth scientists term the *Anthropocene* (Angus 2016) or what some theorists prefer to term the *Capitolocene* (Moore 2016). Riding a tsunami of relatively cheap, readily available fossil fuels, the capitalist thirst for accumulation is creating ramifications on a planetary scale, overseeing exponential rates of deforestation, resource depletion waste creation, and species extinction (Kolbert 2014; Patel and Moore 2018). The specter of a hostile and turbo-charged climate looms large on the horizon, with devastating climatic catastrophes already episodically occurring around the globe. A few examples of climate chaos in recent years include a heat index as high as 163 degrees Fahrenheit (72.8 degrees Centigrade) across the Middle East and Persian Gulf in 2016, recurrent wildfires in California and the western United States, Greenland in 2017, Arctic Sweden in 2018, dying oceans, etc., etc. (Wallace-Wells 2019). A rise of four degrees or more by the century's end would be a world transformed, a less forgiving planet with real existential implications for the humanity and multiple forms of non-human life forms, large and small.

Marx's metabolic analysis can be extended to an examination of global climate change involving the linkage of three dimensions: (1) the impact of capitalism on the global carbon cycle; (2) the Jevons Paradox, in which technological improvements in energy efficiency lead not to decreased but increased utilization of natural resources, and thereby contribute to environmental degradations; and (3) of capitalism in the destruction of carbon sinks (Foster and Clark 2020). As is evident from the imperiled list of planetary boundaries, climate change is not happening in a vacuum but is growing at a time of multiple and other anthropogenic eco-crises, some of which, like ocean acidification and stratospheric ozone depletion, are verging on their own respective tipping points (Wijkman and Rockstroem 2011).

Climate change has been a hot topic – whether in the form of alarm about it or denial of it – for 15 years or more in Australia. As has been the case globally, the

DOI: 10.4324/9781003202530-2

first two decades of the 21st century have been the warmest period since human habitation on the continent. In Australia, four degrees of warming is predicted to dramatically alter rainfall and temperature patterns and distribution, leading to increased incidence and intensity of extreme weather events like droughts, heatwaves, bushfires, floods, and storms (Braganza et al. 2014; Whetton et al. 2014; Reisinger et al. 2014). The impact of climate change in Australia in a Four Degree World is projected to entail the following aspects:

- Temperature increases of 3–5°C in coastal areas and 4–6°C in inland areas;
- Probable declines of precipitation in southern Australia, particularly during winter, of up to approximately 50 percent, but erratic rainfall changes in other regions;
- Substantial increases of possible evaporation of 5–20 percent;
- Snow cover duration disappearing in most alpine regions; and
- Sea-level rise of up to approximately 1.1 meters by 2100, increasing to over seven meters over subsequent centuries even without further global warming (Whetton et al. 2014: 29).

Climate change is having profound impacts on the intensity and frequency of precipitation, heat waves, wildfires, and floods in Australia. These changes are projected to acutely affect Australia's biodiversity, ocean, and agricultural systems, with potentially severe social and economic impacts, particularly in cities, where infrastructures, service provision, and patterns of settlement could face significant disruption (McDonald 2014). Furthermore, Australia will need to grapple with environmental problems within its broader region, including, for instance, food security and eco-refugee migration (Christoff 2014). South-eastern Australia has experienced pronounced warming over the past 50 years, with a 1.1°C increase in maximum temperature and 0.9°C degree increase in minimum temperature (Gergis 2018: 102).

Although it is bordered by three oceans – the Indian, Pacific, and Southern – Australia is often described with good reason as the driest inhabited continent on the planet. Only about 10 percent of Australia's land mass is arable and much of that is marginal when compared with other landmasses. Australia is the developed country most vulnerable to the direct impact from neighboring countries stressed by climate change. Due to both climate change and non-climate-related stresses, the Great Barrier Reef lost 50 percent of its coral cover of the past three decades (Climate Council 2016a: 6). Heat events have become longer, more frequent, and more intense over much of Australia over the past century. Since the 1970s, stints of fire weather in much of Australia have lengthened along with the fire season.

Bushfires have a long history in Australia, but their frequency and intensity worsened in the wake of European invasion and colonization. Collins observes:

There is no doubt that the settlers enormously increased the frequency and spread of fire on the continent, perhaps five- or six-fold. Country that might have burnt in a natural rhythm every 30–50 years before 1788 was being burnt every three years. Certainly Aboriginal people burnt some country on a regular basis but this was focused and low-intensity burning. What becomes clear from a study of fire in European Australia before 1939 is that the settlers introduced a massive dislocation into the established ecological patterns of the landscape. While they would have claimed that they were using controlled fire to fight wildfire, their burning patterns were encouraging the very vegetation that fed increasingly massive conflagrations. It is a tribute to the toughness and staying power of the landscape that is has to some extent survived the European onslaught.

(Collins 2009: 61)

While devastating bushfires occurred over the course the 20th century in various parts of Australia, such as the devastating bushfires that hit Victoria, southeastern South Australia, the Australian Capital Territory, and Tasmania in January 1939, the devastation and frequency of bushfires in Australia have particularly intensified over the past quarter of a century, a period of increasing global warming (Collins 2009).

The Millennium Drought (1997–2009)

What is referred to in Australia as Millennium Drought (1997–2009) proved to be the last major drought of the 20th century and the first one of the 21st century (Gergis 2018: 100). It impacted much of southern and eastern Australia. The average rainfall for southeastern Australia was 512 mm or 12 percent below the 1990–2010 average. Conversely, much of the rest of Australia experienced above-average rainfall, particularly in the northwestern region. The Murray-Darling River Basin, which provides 41 percent of Australia's gross value of agricultural production and 70 percent of water for irrigation, experienced its hottest year on record up to that point in 2007. Parts of the river became mere brown trickles of water. The situation in the Murray-Darling Basin became so grim that the federal government offered farmers who were adversely impacted by the drought a payment of $150,000 to exit their enterprises. Climate scientist Joelle Gergis (2018: 102), formerly based at the University of Melbourne and now at the Australian National University, observes that "[w]hile work in this area is still ongoing, the latest research suggests the Millennium Drought was linked to global warming."

The 2003 bushfires that destroyed nearly 2 million hectares of high country across Canberra, New South Wales, and Victoria constituted a chapter in the Millennium Drought. They were ignited on January 8, 2003, by dry lightning strikes and expanded to over 100 local wildfires and made their way to Canberra on January 14 (Manning 2020: 35–38).

The Victorian bushfires of 2009 proved to be the most devastating bushfire on record to that time in Australia's history. Black Saturday on February 7, 2009, occurred at the tail end of the Millennium Drought and was preceded by an unprecedented heatwave across Victoria and South Australia, translating into Australia's worst bushfire on record to that point (Manning 2020: 50–54). On Black Saturday alone, 173 people perished in bushfires, with over 2,000 homes being burned down and millions of plants and animals being scorched and the town of Marysville in the Yarra Ranges devastated (Gergis 2018: 194–196).

Much of southern Australia experienced unprecedented temperature extremes in the two weeks prior to Black Saturday. Adelaide had nine consecutive days above 35°C and six consecutive days above 40°C. Melbourne had three consecutive days over 43°C, and the temperature broke an all-time record of 46.4°C on Black Saturday, surpassing the previous record of 45.6°C in 1939. Victoria set a record of 48.8°C in Hopetoun.

Kevin Hennessey, a CSIRO climate scientist, in the aftermath of Black Saturday, told the ABC that global warming was contributing to hotter and more intense bushfires:

> The weather system responsible for it was not that different to those that might have occurred . . . in past but the thing that was different was that it was extremely dry, after 12 years of below average rainfalls, and there was extreme heat coming from the interior of the continent. Now whether climate change is responsible for an individual event is very difficult to determine. What we do as climate scientists is look at trends in climate over many decades and what is clear from that is that there has been a warming trend since 1950s and most of that warming trend is very likely due to increases in greenhouse gases.
>
> *(quoted in Manning 2020: 53)*

Ironically, the Victorian Bushfires Royal Commission report initiated by ALP Premier John Brumbey made no mention of climate change possibly having been a driver for the devastating events (Manning 2020: 51). Nonetheless,

> [e]nvironmental scientist Professor Peter A. Gell of Ballarat University pointed out in a submission to the royal commission that "South-east Australia is a global climate hot-spot", that "extended drought creates the highest risk of catastrophic wildfire", and that "very extreme and catastrophic conditions will occur more frequently."
>
> *(Collins 2009: xxvii)*

The Millennium Drought resulted in the implementation of water restrictions in most Australian major cities and in increased electricity prices (Van Dijk 2013: 1040).

Catastrophic rains and floods

The Millennium Drought finally broke in 2010–2011, however, and during this period Australia experienced huge floods associated with the strongest La Nina event since 1917. The six months from July to December 2010 were Australia's wettest ever (Manning 2020: 127). Heavy rainfall in Queensland resulted in extensive flooding, adversely impacting 78 percent of the state and some 2.5 million people, with 33 people dying and 29,000 homes and buildings being inundated (Gergis 2018: 203). On January 10, 2011, the inland country town of Toowoomba, Queensland, situated at 700 meters above sea level, experienced heavy flooding, with 138 millimeters of rain falling between 1 and 2 pm. Tropical cyclone Yasi struck Mission Beach on February 3, 2011, and, rated as a category 5 event, entailed speeds of up to 285 kilometers per hour and a five-meter tidal surge. Reportedly Yasi inflicted an estimated $1.6 billion and $600 million damage to the agricultural and tourism industries, respectively (Climate Council 2020: 6).

Tropical cyclones impact both adversely and positively upon coral reefs, such as the Great Barrier Reef in Australia, which has undergone three mass bleaching events during the period 2016–2020. The Climate Council reports:

> For example, the southern section of the Great Barrier Reef was spared from mass bleaching in 2016 because of late summer cooling of ocean temperatures from ex-Tropical Cyclones Winston and Tatiana. Whereas, on the downside, climate change could result in more intense tropical cyclones hitting the Australia coast, which will in turn increase the risk of physical damage to reefs. In 2016, Tropical Cyclone Yasi passed over large areas of the Great Barrier Reef. Coral damage was reported across an area of approximately 89,000 km² of the Great Barrier Reef Park. In total 15% of the park sustained some damage and 6% was severely damaged.
>
> *(Climate Council 2020: 5)*

During La Nina years, generally about twice of as many tropical cyclones hit Australia compared to El Nino years.

The Angry Summer, 2012–2013

The Angry Summer of 2012–2013 included Australia's hottest day, week, month, and year averaged across the country, despite occurring without an El Nino event (Gergis 2018: 184). The heatwave lasting from December 25, 2012, to January 18, 2013, constituted the hottest heatwave on record in Australia and encompassed 70 percent of the continent, with 123 temperature records being broken across every state and territory. The national daily average temperature rose to an unprecedented high of 40.3°C on January 7, 2013. Meteorologists added two new color categories to Australia's weather prediction maps to accommodate the scale of the exceptional heat.

Despite being a huge country in area, most of which is sparsely populated, Australia is a highly urbanized country with some 60 percent of Australians living in a capital city, namely Adelaide, Brisbane, Canberra, Darwin, Hobart, Melbourne, Perth, and Sydney, and almost 90 percent of Australian living in cities and towns

TABLE 1.1 Projected Increases in Days over 35°C for All Australian Capital Cities and Selected Other Cities between 2015 and 2090

City	2015	2030 RCP4.5[a]	2030 RCP4.5[a]	2090 RCP8.5[b]
Adelaide	20	26	32	47
Alice Springs	94	113	133	168
Brisbane	12	18	27	55
Cairns	3	5.5	11	48
Canberra	7.1	12	17	29
Darwin	11	43	111	265
Hobart	1.6	2	2.6	4.2
Melbourne	11	13	16	24
Perth	28	36	43	63
Sydney	3.1	4.3	6	11

RCP4.5[a] refers to an intermediate emissions scenario
RCP8.5[b] refers to a high emissions scenario
Source: Commonwealth Science and Industrial Resource Organisation (2015).

numbering over 1,000 people. Australia's capital cities have been in recent decades experiencing longer, hotter, and more frequent heatwaves. Melbourne is now on average 1.5°C hotter than it was during the period 1950–1980 and Adelaide has nearly doubled its heatwave days since then (Gergis 2018: 185). Table 1.1 depicts projected increases in temperature over 35°C for all Australian capital cities and selected other cities between 2015 and 2090. As summer in most Australian capital cities becomes hotter, many people will try to beat the heat by turning on air-conditioners in their homes, offices, schools, hospitals, shops, restaurants, shops, and motor vehicles, a measure that will result in more greenhouse gas emissions, particularly if the source of electricity is coal-fired power plants.

Table 1.1 depicts projected increases in days over 35°C for all Australian capital cities and selected other cities between 2015 and 2090.

During winters, snowfalls in Australian Alps have been diminishing, with the snow season shortening and less meltwaters, thus adversely impacting agricultural productivity and hydropower generation (Gergis 2018: 192). CSIRO modelling forecasts that the Australian ski season could shrink up to 80 days a year by 2050 under worse-case predictions for climate change and a low-emissions scenario forecasts that the snow season will become 20–55 days shorter across the Australian Alps in Victoria and New South Wales.

The 2019–2020 mega fire

The 2019–2020 bushfires began in the wake of Australia's hottest summer on record. They commenced in August, and by September 10, some 80 bushfires were burning in Queensland and some 60 bushfires in New South Wales. These occurred in absence of El Nino conditions typically associated with severe bushfires in past years. Sydney was covered in smoke from early December until January 2020.

A tipping point of sorts began in spring 2019 when large portions of Queensland and New South Wales experienced high temperatures and bushfires that contributed to Sydney as Australia's largest and most visited city being enveloped with smoke. While this climate crisis impacted his Sydney electorate, Coalition Prime Minister Scott Morrison jumped ship with his family for a holiday in Hawaii in mid-December 2019, texting Australia Labor Party opposition leader Anthony Albanese that he would be gone for a week and that Michael McCormack, the Nationals leader and deputy prime minister, would be at the helm. Many Australians became outraged by the prime minister's action, particularly after two firefighters lost their lives fighting bushfires in New South Wales. Sensing this outrage, Morrison and his family returned to Sydney and he publicly apologized for having offended so many Australians by his actions.

When in early January 2020 Morrison visited Cobargo, a small country town in New South Wales, he encountered overt hostility from various residents whose community had been devastated by a bushfire. Despite considerable political fallout, Morrison insisted that Australia's present emissions reduction policies are adequate to mitigate a climate crisis and bushfire risks while bushfires burned in six states, including Tasmania. Due to the bushfires, Australia's greenhouse gas emissions had increased dramatically. On December 26, bushfires were burning in the Stirling Ranges of Western Australia and by December 30, there were three bushfires in East Gippsland, Victoria. On December 31, the bushfire reached Mallacoota, Victoria, prompting over 4,000 people, both residents and tourists, to find sanctuary of sorts on the beach. They were evacuated by the Australian Navy on January 3.

Like Sydney, although not as often, Canberra also experienced smoke from the bushfires, to the point that government offices, universities, and other facilities were closed for 48 hours because the air had become extremely dangerous in the first days of the New Year. During this period, the national capital was reportedly the most polluted city in the world. Smoke from the Australian bushfires even crossed the Tasman Sea to New Zealand and even reached as far away as Indonesia. Melbourne residents woke up to a thick haze on the morning of January 5, 2020, not so much due to smoke from the bushfires in New South Wales and East Gippsland on previous days, but largely from bushfires in Tasmania across the Bass Strait. Canberra experienced the worst air quality of city on world on New Year's Day 2020 (Manning 2020: 205). It recorded the worst air quality reading again on January 5 and 6, 2020, prompting the authorities to shut government offices, businesses, and the Australian National University. On January 20, a freak hailstorm hit Canberra and the adjacent sections of New South Wales and even parts of Victoria, inflicting extensive damage.

Melbourne was next to experience the world's worst air quality on a given day. Penrith, a town west of Sydney, recorded a temperature of 48.9°C on January 6, making it the hottest site on the planet on that day. The 2019–2020 mega fire burned in six states at one point or other and "burned 186,000 square kilometres, 21% of Australia's broad-leaf temperate forested area" (Flannery 2020: 57). The 2019–2020 mega fire resulted in the deaths of over 80,000 livestock, particularly dairy cattle, adversely affected some 9 million sheep and 2 million cattle (Manning 2020: 232). They released roughly 900 million metric tons of carbon dioxide.

Of the mega-fire of 2019–2020, Flannery observes:

> The megafires that raged for months that summer transfixed the world. They also woke Australia to the dangers the nation is sleepwalking into. It's now clear that a signature event of the Anthropocene will be the climate-change-driven megafire. Other continents have seen megafires before with blazes in recent years devastating California, the Amazon, Southern Europe and Siberia. But Australia's megafire was like nothing previously recorded on Earth.
>
> *(Flannery 2020: 57)*

By February 7–10, two-thirds of the bushfires in New South Wales had been extinguished, partly due to heavy rains and flash floods along the coast. On February 8, Cyclone Damien struck the Dampier-Karratha region of Western Australia. The 2019–2020 bushfires destroyed 5,900 buildings, including 2,779 houses, and wiped out the tourist season for some coastal towns (Flannery 2020: 63).

The impact of climate change on Indigenous Australians

Of all Australians, Indigenous people, particularly those residing in remote communities in central and western Australia, are the most vulnerable to rising temperatures and other climatic changes, such as drought and flooding. Roughly three-quarters of Australia's Indigenous people reside in cities or regional towns with the remaining quarter residing in remote areas.

Anthropologist Donna Green conducted research on how climate change might affect remote Indigenous Australian communities and how they plan to respond to it. She convened a workshop that assembled 30 Elders and Traditional Owners of the Land from across northern Australia and 30 researchers and scientists (Green 2009: 220). The group decided to prioritize the impact of climate change on Torres Strait Islander peoples who reside on islands situated between mainland Australia and Papua New Guinea. The Islanders reported in recent years, the King Tides seemed higher and more powerful. The Islanders were also concerned about how water inundations might result in contamination of freshwater supplies or flood their landfill rubbish dumps.

Marcus Barber (2011), an anthropologist who works for the Commonwealth Scientific and Industrial Research Organisation (CSIRO), notes that the Garnaut Review on climate change commissioned by the Australian government predicts that without significant mitigation, Darwin, the capital of the Northern Territory, will have over 200 days each year above 35°C by 2070. He has spent a significant portion of his time working at CSIRO travelling to remote northern Australia, speaking to Indigenous people about issues of water and environmental change. His Indigenous informants increasingly referred to "climate change" and "global warming" without prompting, suggesting the diffusion of climate science concepts to even remote areas of the planet. Barber (2011) asserts that one challenge for anthropologists is how to find appropriate means of conversing with people who are removed from the generators of climate change. Indigenous Australians across Arnhem Land in the Northern Territory are becoming increasingly aware of anthropogenic climate change due to reports in mass media, reports writing by Indigenous rangers, and governmental environment reports. As a result, the Yolngu now regard climate change as a major threat to their way of life.

The Kimberley region of northwestern Western Australia faces an increase in extreme heat days (days for 35°C), with CSIRO and the Bureau of Meteorology projecting an "increase from an average of 81 days above 35 degrees, to up to 195 days by 2050 and potentially up to 286 days by 2090" (Quickie 2019: 1). Indigenous people constitute about 46 percent of the Kimberley population and primarily reside in rural or remote communities (Quickie 2019: 37).

Overall, the threat of adverse impacts emanating from climate change is growing among Indigenous people. In the words of Melanie Koolmatrie of the Ngarrindjeri community, reflecting on the changes she has witnessed:

> We have always lived by the water. We are water people. We are water spirits. I only have one country and it is this country. If that goes, I have no other home. . . . It is very hurtful. Our land is sick. Something needs to be done. It needs to be done now.
>
> *(Hero Project 2009)*

Green, Billy, and Tapim (2010) maintain that Indigenous Australians have accumulated rich ecological knowledge, including climatic and weather observations, which potentially can fill gaps in the climate data on seasonal changes for tropical northern Australia. Indeed, the Australian Bureau of Meteorology operates an online Indigenous Weather Knowledge Project, which makes three weather calendars available for Indigenous communities. Green, Billy, and Tapim (2010: 351) maintain that "Indigenous knowledge may well be one of the keys in understanding how best to engage in culturally appropriate climate change adaptations for these communities."

Unless greenhouse gas emissions are drastically reduced both globally and in Australia, Indigenous people, particularly in remote areas, face enormous challenges to their physical, mental, and socio-cultural well-being. Joelle Gergis (2018: 220) paints a grim picture for particularly inland Australia in terms of rising average temperatures by 2100 unless drastic reductions in emissions were to occur in the interim:

> The biggest changes will affect inland Australia, where up to 5.3°C of warming is projected under high-emissions scenarios. In a town like Alice Springs, this means the number of days over 40°C will rise from seventeen per year to an average of eight-three per year, close to a five-fold increase. It's very likely

that parts of the outback will recover summer temperatures of over 50°C, making much of inland Australia incredibly unlivable.

(Gergis 2018: 220)

Given this grim reality, Green and Minchin aptly assert:

Without adaptive planning and action, climate change poses a particular threat to Indigenous Australians' sense of well-being (alternatively referred to as social and emotional well-being, or psychosocial health). If country becomes "sick" through climate impacts, environmental degradation, or the traditional owners' inability to fulfill cultural activities to care for country, people have reported how they feel this "sickness" themselves.

(Green and Minchin 2014: 267–268)

Nash et al. (2018) conducted a series of interviews on the perceptions of climate change of Indigenous people in various communities in the Upper Georgina River Basin in Northwest Queensland, including Mount Isa. While not all of their informants "are completely engaged with or convinced by the scientific discourse of climate change, they have observed the gradual transformations and shifts in local ecosystems that such changes bring about" (Nash et al. 2018, p. 117).

Drawing upon focus group discussions with Indigenous people from multiple locations in north-east Queensland, Alice Springs, and Adelaide, Nursey-Bray et al. (2019) found that their informants articulated their fear of how increasing temperatures might impact upon well-being due to inadequate housing and lack of air-conditioning, mediated by possible pre-existing health problems in both rural and urban areas. Just as was the case for hundreds of Australians in southeastern Australia due to the massive bushfires of 2019–2020, Indigenous Australians face the tragic possibility over the course of next several decades of becoming internal climate refugees.

Conclusion

In their most recent annual "state of the climate" report, the Commonwealth Science, and Industrial Research Organisation and the Australian Bureaucracy of Meteorology (2020) reported that Australia's climate has warmed an average of 1.44°C since national records commenced in 1910. Furthermore, 2019 was Australia's year on record. The seven years from 2013 to 2019 all ranked in the nine warmest years on record for Australia. A decline of about 16 percent in April to October rainfall in southwest Australia has occurred in Australia since 1970. The May–July rainfall across the continent has decreased about 20 percent since 1970. Furthermore, a decline of about 12 percent in April to October rainfall has occurred since the late 1990s. Rainfall has increased across parts of northern Australia since the 1970s. Northern Australia has experienced a wetter climate across all seasons, particularly in the northwest and during the months of October to April. Extreme fire weather and the length of the fire season has occurred across large parts of Australia since the 1950s, particularly in the southern portion of the continent. Although the number of tropical cyclones in Australia has decreased since 1982, their intensity has heightened. Furthermore, the oceans, namely the Pacific, Southern, and Indian, have been acidifying and have warmed about 1°C since 1910, resulting in longer and more frequent marine heatwaves. Sea levels have risen around Australia, resulting in inundation and damage to coastal infrastructure and communities. The warming of the seas and duration of marine heatwaves threatened the resilience of coral reef ecosystems.

The report forecasts the following climatic developments in Australia over the course of coming decades:

- Continuing increases in air temperature, with more heat extremes and fewer cold extremes;
- Ongoing decline in cool season rainfall across many regions of southern and eastern Australia, contributing to longer droughts, accompanied by more intense, short duration heavy rainfall events;
- An increase in the number of dangerous fire weather days for southern and eastern Australia;
- Ongoing sea level rise, warming, and acidification of the oceans surrounding the country; and
- Fewer tropical cyclones, but a larger proportion being highly intense.

Gergis (2018: 220) presents a sobering climatic scenario for Australia by 2100, noting: "While there are differences in the climate change projections for each Australian region, the average temperature increase across the country will be 4°C by the end of the century." Indirect effects of climate change on Australia include polluted river systems, the disappearance of bees and butterflies in many areas, a reduction in the numbers of birds, and hazardous air quality in cities (Jones 2020: 141). The Climate Council (2019b) projects that the Australian property market may lose $571 billion in value by 2030 due to climate change and extreme weather events. Over $226 billion in commercial, industrial, road, rail, and residential assets will be at risk due to sea level rise alone by 2100 if greenhouse gas emissions continue to rise at high levels. Furthermore, climate change-related events may result in reduced agricultural productivity and labor productivity, projected to exceed $19 billion by 2030, $211 billion by 2050, and $4 trillion by 2100.

While hoping for a world which is 2°C or less, an objective aimed at in the UN Framework Conference on Climate Change Paris Agreement of 2015, Palutikof, Barnett, and Guttart (2014: 216) argue that "humanity should certainly plan for more, including average warming of 4°C by the end of this century. . . . A Four Degree World." In terms of Australia, they envision the following three scenarios for the 2070s:

- *Terror Australis*
  - A minimalist and pro-growth state
  - Public policy being hostage to corporate or capitalist interests

- Climate change adaptation oriented toward economic growth through "strategies such as the progressive privatization of essential services and public goods."

- *Terrior Australis*

  - A social democratic state
  - Public policy achieved "through between the state, businesses, organized labour and localities"
  - Emphasis on equity and liberty

- *New Atlantis*

  - A purposeful state
  - Public policy is oriented toward a "convergence between the state, capital, labour and civil society"
  - Emphasis on justice, the good life, and global citizenship (Palutikof, Barnett, and Guttart (2014: 220–230).

Of three future scenarios laid out by Palutikof, my preference is clearly for *New Atlantis*. My eco-socialist vision for Australia goes beyond even this scenario in that it calls for transcending capitalism as part of a larger global effort, an issue that I lay out in greater detail later in this book.

2
AUSTRALIAN CAPITALISM AS A DRIVER OF CLIMATE CHANGE

Norwegian anthropologist Thomas Hylland Eriksen (2016) argues that the world system has become *overheated* on numerous fronts, including energy consumption, mobility of people hither and thither across their respective countries and the globe as a whole, urbanization, production of waste, and information overload stimulated by digitalization. He aptly applies the overheating metaphor to Australia:

> A major exporter of coal and iron ore, but also a country whose inhabitants drive, fly and own air-conditioners, Australia ranks near or at the top of any list of CO_2 emitters. At the same time, the bleaching of corals and the increasing frequency of bushfires, floods and droughts that Australia has witnessed in recent years also make it one of the prime victims of climate change.
>
> *(Eriksen 2016: 55)*

While some Australian corporations, such as the BHP Billiton, once referred to "The Big Australian," started out on home soil, most multinational corporations operating in the country are largely foreign owned, including by ones based in the UK, the United States, Japan, and increasingly over the past decade or so China. This includes the former BHP Billiton, renamed simply BHP.

Australia has been described as the 51st state because it has become Americanized in many ways, particularly since World War II, but it retains its own distinct identity. This identity is colored by its parliamentary system with the Queen of England as its head of state. Also, compared to the United States, Australia has a stronger (if eroding) welfare system, greater emphasis on multiculturalism and greater attention to Indigenous peoples, greater secularism as manifested by electing some self-proclaimed atheists as prime ministers, and more subdued patriotism. Although this take is undercut somewhat by the strengthening of the US–Australian

DOI: 10.4324/9781003202530-3

military alliance in recent years, particularly since 9/11, Australian political scientist Dennis Altman basically correctly observes:

> Australia is not, as the United States has been since World War II, a militarised society. We have a far smaller military establishment and far less employment depends upon military expenditures.
>
> *(Altman 2006: 20–21)*

From a world system theory perspective, Australia may now seem to be what's called a minor core country. In this theory of unequal exchange, core or developed capitalist countries exploit the peripheral or the poorest countries as well as the semi-peripheral countries (Shannon 1996; Wallerstein 2004). The periphery is exploited by both the core and the semi-periphery – as, for instance, semi-peripheral giant China now exploits various African countries. In fact, Australia isn't a core power by the definition implied in world system theory, but it might be more appropriately termed an advanced semi-peripheral country given the strong role of foreign capital within its borders. Nearly four decades ago, Crough and Wheelwright (1982: 1) referred to Australia as a *client state* of international capital because by this time Australia had the "highest level of foreign ownership and control of all the advanced countries of the world except Canada, which is now making a serious attempt to buy back its energy sector."

Australia as a quarry nation

Numerous commentators have characterized Australia as a "quarry nation" that may eventually exhaust mineral resources that have contributed to its wealth. In Australia mineral resources below the ground constitute Crown land and must be leased by mining companies to extract mineral resources, such as iron ore, bauxite, uranium, and coal. State governments exert jurisdiction over regional and local matters such as minerals and energy, biophysical resources, agriculture, and infrastructure, including airports, transport, and urban planning. They have control over mining leases and exploration permits. State and Territory governments levy royalties on mineral commodities, including coal, although royalty rates vary. Although the Australian Constitution does not grant the federal government jurisdiction per se over mining operations, it can override state and territorial policies on mining operations if it deems them to be inconsistent with federal legislation.

Crough and Wheelwright (1983) delineate four features of the Australian mineral industry that have existed since at least the late 1960s: (1) a high level of foreign ownership, (2) large inflows and outflows of industrial capital, (3) an export-import orientation, and (4) heavy reliance on imported equipment rather than domestically manufactured equipment. They assert that under this scenario, Australia has become a *client state* "whose main function is to shape the future development of the economy in such a way that profits of foreign corporations have first priority, and the needs of the Australian people the least priority" (Crough and Wheelwright

1983: 35). In 2009, BHP and Rio Tinto were 76 percent and 83 percent foreign-owned, respectively (North 2016: 169).

While corporations invariably resist government intervention if it entails regulation of their labor and environmental practices or profits, they want state involvement in situations where it benefits them. As Corrighan observes,

> Capital, and particularly multinational capital, *demands* that the state provide sufficient inducements before investment and expansion will be undertaken. This applies especially to mining capital, which is certainly *not* opposed to the state having a major role in development, as long as this role is confined to such areas such as massive capital outlays for the provision of infrastructure, which capital itself will not provide.
>
> *(Corrighan 1980: 30)*

The mining industry repeatedly asserts that it is a significant contributor to Australia's economic prosperity. In actual fact, as Denniss reports:

> The taxes and royalties paid by all mining companies to all levels of government account for less than 5 per cent of government revenue. And that's the total amount paid, not the net amount after we deduct the subsidies they receive. The Queensland government earns more from speeding fines and car registrations than it does from the coal industry.
>
> *(Denniss 2016: 32)*

Prior to the 1970s or thereabouts, resource-based development tended to be viewed in a positive manner and given generous subsidies by Australian states, particularly in Queensland and Western Australia, not only because it created profits for mining companies but also because it created royalties, tax revenue, and employment. Conversely, as Bruckner et al. (2014) observe, mining inflicts risks upon Australia's environmental, social, and economic sustainability, ones that tend to be downplayed by the corporate and state sectors. While the notion of *resource curse* has been applied primarily to developing societies because much of the generated wealth leaves them in the form of profits and mineral resources, benefiting primarily core countries and resulting in environmental damage within the country itself, various scholars have observed that the resource curse can also have an impact on developed societies, such as Australia (Goodman and Worth 2008). Coal production and export in particular create a relatively a small number of jobs in Australia (Davidson and de Silva 2013) and contribute to several ecological and climatic crises both domestically and globally.

Since the mid-1980s, under the framework of neoliberalism or economic rationalism, both Labor and Coalition governments at federal and state levels have subsidized and promoted the exploitation of Australia's coal and natural gas supplies. An Australia Institute study revealed that the Australian federal and state governments provided a total of $10.3 billion of subsidies and tax breaks to the fossil fuel industry in 2020–2021 (Campbell, Littleton, and Armistead 2021).

The state's willingness to spend billions (an estimated $10 billion in 2015) to construct the railroads, coal ports, and much of the other infrastructure needed by the mining industry demotes Australia to semi-peripheral status (Lowe 2016: 35). Since World War II, the Australian resource sector has developed strong ties with Japan based on coal exports and, even more recently, with China based on exports of iron ore but increasingly also of coal (Cleary 2011; Pearse, McKnight, and Burton 2013). Australia, then, is feeding the beast instead of harnessing it.

Most sectors of the Australian economy, including the mining industry, are dominated by foreign investment, particularly on the part of Anglo-American corporations. Over 80 percent of the mining sector is foreign-owned (Paul 2012: 8). The resource sector is the largest component of the Australian economy. Mining and energy exports increased 29 percent of merchandise exports in 1993–1994 to 44 percent in 2006–2007 and more than 60 percent in 2009, mostly to East Asia (Paul 2012: 42). Australia is the world's leading producer of iron ore, close to 40 percent of the world total, and the leading exporter, nearly 60 percent, of iron ore (Garnaut 2019: 115). Roughly 70 percent of Australia's exported iron ore goes to China, where it is used in steel production, about half of the global output of steel, which results in carbon dioxide emissions and has assisted China in becoming the factory of the world and contributed to its dubious status as the leading greenhouse gas emitter globally.

Australian governments at both the federal and state levels have traditionally subsidized the costs to private enterprise of large-scale development projects and worked in close concert with corporations, including mining companies. Large state government investments in electricity generation and ports and harbor developments function often as a subsidy to the private sector to encourage mineral exports, including coal. Mining companies have played a key role in Australian politics and have been involved in the ouster of two Labor governments, namely the Whitlam government in 1977 and the Rudd government in 2010. Conservative forces, including those in the mining industry, contributed to the ouster of Prime Minister Gough Whitlam because of his government's plan to nationalize energy resources and its interference with US military and intelligence operations based in Australia (Paul 2012).

Coal mining in Australia

Australia is part of a global political economy highly dependent on coal production and consumption. While from a world system theory perspective, one might define Australia as a minor core country, the willingness of the Australian state to construct much of the infrastructure needed to facilitate the operations of the coal industry, such as railways and coal ports, demotes Australia to semi-peripheral status. In the case of Australia, Europeans discovered black coal in the Newcastle, New South Wales, area in 1791 and began exporting it in 1799. These early coal mining activities made a significant contribution to the progress of European set-tlement in Australia. The progressive spread of settlement to other locations in Australia led to further discoveries and mining of black coal including the discovery of coal near Ipswich in 1825, at Cape Paterson in Victoria in 1826, and Irwin River in Western Australia in 1846. Knowledge of the existence of brown coal in Victoria dates back to 1857. The Yallourn North open-cut coal mine began production in 1889. Over 200 years later about 9,100 million tons of black coal and about 2,300 million tons of brown coal have been mined and the Australian coal industry provides significant employment, capital investment, and domestic and export income to the national economy.

While black coal is found in South Australia, Western Australia, and Tasmania, New South Wales (23 percent) and Queensland (63 percent) have the largest share of Australia's coal deposits (Geoscience 2020: 1). Brown coal reserves exist in South Australia, Western Australia, Tasmania, Queensland, and Victoria. Brown coal is only mined in Gippsland in eastern Victoria which is used to power the three remaining coal-fired power plants. Almost 80 percent of Australian coal mined comes from open-pit mines and most of them may be used with little or no processing.

Since the mid-1980s, under the framework of neoliberalism or economic rationalism, both Labor and Coalition governments at federal and state levels have subsidized and promoted the exploitation of Australia's coal and natural gas supplies. As a modest favor in return, the fossil fuel industry routinely makes financial contributions to the major political parties. Over the period of 2013–2015, the fossil fuel industry contributed $2,365,250 to the Liberal Party, $221,787 to the National Party, and $1,108,528 to the ALP (North 2016: 204).

The coal industry keeps expanding with governmental support. Both Coalition (Liberal and National parties) and Australian Labor Party (ALP) governments at both the federal and state levels have consistently subsidized the exploitation of Australia's coal and natural gas supplies and more recently coal seam gas supplies and supported the expansion of coal exports. The Green House Mafia, a term that leaders of a consortium of industries involved in the production of fossil fuels and coal-based electricity generation applied to themselves during the tenure of Prime Minister John Howard, "has dominated almost every greenhouse-related consultative committee established by the federal government and its agencies" (Pearse 2009: 41). Just as this was the case for the federal ALP governments under Prime Ministers Kevin Rudd, Julia Gillard, and again Kevin Rudd between 2007 and 2013, it has been the case in subsequent Coalition governments. Despite his insistence that climate change was the greatest moral challenge of this generation, former Prime Minister Kevin Rudd announced on December 12, 2008, that $AU580 million of public money would be allocated to the expansion of coal facilities in the Hunter Valley and the Port of Newcastle (Jowit and Wintour 2008: 1). Even while preparing its Carbon Pollution Reduction Scheme (CPRS) legislation, an emissions trading scheme, which never was passed, the Rudd government poured huge subsidies into coal-fired power plants.

While Australia, with some 25.7 million people, contains only 0.03 percent of the world population, it produces roughly 1.2 percent of global greenhouse emissions, is the 14th largest emitter globally, has the highest emissions per capita of any OECD country, and is only behind various small petro-states, such as Qatar, on this count; is the largest coal exporter and the largest liquid natural gas (LNG) exporter; and is the fifth-largest miner of fossil fuels, only superseded by China, the United States, Russia, and Saudi Arabia (Swann 2019: 1).

Coal constitutes Australia's second-largest resource export commodity, only superseded by iron ore exports. In terms of coal exports, roughly half is thermal coal and half coking coal, with thermal coal being mined primarily in New South Wales and coking coal primarily in Queensland (Swann 2019: 10). With three-fourths of its black coal production exported, Australia was the world's leading coal-exporter until 2011, when it slipped to number two behind Indonesia for a few years, later regaining its number one status. In 2015, 32 percent of Australia's black coal exports went to Japan, 17 percent to China, 15 percent to South Korea, and 13 percent to India (International Energy Agency 2018: 176).

Table 2.1 depicts the five leading producers of coal worldwide, with Australia ranking number four.

In 2016, China produced 3,243 megatons of coal, compared to 503 megatons in Australia (International Energy Agency 2018). In 2017, however, Australia was the largest coal exporter in the world, having exported $US40.6 billion worth of coal (Cunningham, Nanuffeliin, and Chambers (2019)). Australia exports about half of the coal that it mines in exchange for gross revenue of some $45 billion per annum, with state governments receiving 10 percent royalties from this amount (Orgill 2013: 114). Australia produced less than 100 million tons of oil equivalent (toe) of coal per annum during 1980s, a figure which increased to 150 million toe per annum by 1996; and 250 million toe per annum by 2010 (Brett 2020: 18). Following a short blip during the Global Financial Crisis of 2008–2009, coal production increased to about 300 million toe per annum by 2013, a level which has stayed steady since then.

While Australia has numerous companies involved in coal mining, the Big Four players are BHP, Rio Tinto Australia, Glencore, and Anglo American, which are all primarily foreign-owned and operated. Coal constitutes a significant source of Australia's domestic energy consumption. It relies upon both black and the even

TABLE 2.1 World Production of Coal

Rank	Country	Black coal (Mt)	Brown coal (Mt)	Total coal (Mt)	Percentage (%)
1	China	2344	1127	3471	46
2	USA	917	73	991	13
3	India	539	41	580	8
4	Australia	345	70	414	5
5	Russia	256	78	334	4

Source: Adapted from Geoscience (2020: 4).

dirtier brown coal for electricity production but also upon natural gas, hydroelectricity, and small amounts from wind or solar sources. In terms of the percentage of electricity generated from coal, Australia is in fifth place internationally at 69 percent. Coking coal is used for steel production and by-products of coke-making include coal tar, ammonia, light oils, and coal gas. While much Australian coal in the past had been exported to Japan, South Korea, and Taiwan, more and more Australian coal, including coking coal which is used in steel production, goes to China.

Sinclair Davidson and Ashton de Silva (2013) make a distinction between the *coal extraction industry* and the *coal economy*. They assert that "for every job created in coal mining 3.7 jobs are created nationally in the Australian economy making up a direct and indirect workforce of over 180,000 people in 2011–12" (Davidson and de Silva 2013: 1). Relying in part upon Australian Bureau of Statistics data, Davidson and de Silva (2013: 2) report the following gross value contributions of the coal industry, the coal economy, and the resource economy, respectively, in terms of "total output, supply and demand side effects, to the overall Australian economy in 2011–12: coal industry 1.8 percent, coal economy 4.2 percent, and resource economy 23.7 percent." They reported that in 2011–12 there were 49,000 employees in the coal industry and 181,200 employees in the coal economy. Davidson and de Silva also list, indicated in Table 2.2, the top ten Australian industries reliant on coal.

Most of Australia's coal is mined in the Bowen Basin in Queensland and the Hunter Valley in New South Wales, but there are also coal mines in Western Australia south of Perth and lower grade brown coal is mined in Victoria. Nine terminals at seven ports stretching along the east coast of Australia serve as the coal exporting sites, with Newcastle and Gladstone the largest, both competing with Richards Bay in South Africa for the distinction of the world's largest coal port.

Australia accounted for about 8 percent of the world's brown coal production and ranked fifth largest producer after Germany (20 percent), China (11 percent),

TABLE 2.2 Top Ten Industries Dependent on Coal*

Industries highly dependent on coal	Production of coal required ($AU)
Electricity generation	20,837
Iron and steel manufacturing	6,835
Non-ferrous metal ore mining	3,239
Iron ore mining	2,611
Coal mining	1,466
Cement, lime, and ready-mixed concrete manufacturing	944
Non-metallic mineral mining	664
Pulp, paper, and paperboard manufacturing	503
Basic non-ferrous metal manufacturing	328
Grain mill and cereal product manufacturing	268
Oil and gas extraction	221

*Adapted from Davidson and de Silva (2013: 6).

USA (9 percent), Russia (9 percent), and Greece (8 percent). Australia has surpassed Indonesia in coal export, making it once again the largest coal exporter in the world (Swann 2019: 28).

Australia's largest destination for thermal coal exports is Japan, accounting for about 40 percent in 2018, although Japan's share has declined by about half compared to a decade ago (Cunningham, Van Uffelen, and Chambers 2019: 32). Conversely thermal coal exports to China have increased from less than 2 percent in 2008 to about one-quarter today. India, China, and Japan account for about two-thirds of Australia's coking or metallurgical coal exports.

Gautam Adani is an Indian multi-billionaire who seeks to build the largest coal mine in Australia's history in central Queensland's Galilee Basin. He owns many companies in India, including ports, property and oil and natural gas exploration schemes, and coal-powered electricity plants, and is well connected with Prime Minister Modi (Krien 2017: 14). Nearly 60 percent of India's electricity supply comes from coal-fired power plants and the government seeks to double coal production by 2020, utilizing both domestic and imported Australian coal mined by Adani (Talukdar 2016). The Coalition initially made a commitment to provide a subsidy of $1 billion so that Adani can construct a rail line from the mine to Abbott Point, where the coal would be shipped to India, but later backed off on the offer.

Petroleum, natural gas, and coal seam gas industries

Compared to large petroleum producers such as Saudi Arabia, Venezuela, the United States, Russia, Norway, and other countries, Australia is a bit player, even more so than in the past. Since 1973 when the International Energy Agency began recording petroleum production levels, Australia's domestic crude petroleum production declined 23 percent between 2006 and 2016 to 17 million tons (Mt), its lowest recorded level (International Energy Agency 2018: 42). About 80 percent of the crude petroleum produced in Australia is exported, mainly to Thailand, Singapore, and China. The Australian Institute of Petroleum (AIP) functions as the main consortium of domestic petroleum production in the country, consisting of 31 companies, with the four largest operators being BP, Caltex, Mobil, and Viva (International Energy Agency 2018: 49). Conversely, in 2016, "Malaysia was the largest crude oil exporter to Australia, accounting for 28% of Australia's total imports, followed by the United Arab Emirates (17%), Indonesia (12%), and Vietnam (5%)" (International Energy Agency 2018: 43). Australia has four main pipelines for transporting petroleum and petroleum products domestically.

Western Australia constitutes the primary locus of the Australian petroleum industry, largely based on the development of the reserves of the North West Shelf and other hydrocarbon basins from which it extracts crude oil and natural gas. Petroleum production in Australia increased after 1980, peaked in 2000 at 805,000 barrels per day, fell to 630,522 barrels per day in 2003, and further to 562,000 barrels per day in 2006 (U.S. Energy Information Administration 2007). Woodside Petroleum, Santo, BHP, Apache Energy, BP, Chevron, ExxonMobil, Shell, Ipex, and ConocoPhillips are operators in the Australian petroleum industry. Australian petroleum is both used for domestic purposes and exported to Asian countries.

Table 2.3 depicts the production of petroleum liquids by state/territory in 2019. Table 2.4 depicts the production of natural gas by state/territory in 2019.

Woodside discovered huge deposits of natural gas, mostly methane, off the northwest coast of Western Australia in the 1980s. It liquefied, stored, and transported natural gas with new technologies and exported its first cargo of northwest shelf LNG to Japan in 1989 (Brett 2020: 19). Woodside had shipped some 3,000 shiploads of LNG to Japan by 2012, making Australia the world's the third-largest LNG exporter by 2013. Bechtel began to ship LNG from Curtis Island off the coast north of Gladstone to Asian customers by 2015, thus enabling Australia to become the world's largest LNG exporter. With so much LNG being exported, Australians found themselves paying the highest gas prices in the world by 2017. Reucassel reports:

> The natural gas exporters – Santos, Shell, ConocoPhillips – entered into long-term contracts overseas, which meant domestic prices suffered and we all, domestic and industry started to pay a lot more for our gas. In 2014, gas

TABLE 2.3 Production of Petroleum Liquids by State/Territory (Millions of Barrels), 2019

State/Territory	Crude	Condensate	Liquid Petroleum Gas
New South Wales	–	–	–
Northern Territory	0.1	20.7	13.9
Queensland	–	–	–
South Australia	12.5	1.8	2.6
Tasmania	–	0.4	0.4
Victoria	3.3	7.5	7.8
Western Australia	31.6	63.3	6.1
Total	47.6	93.7	30.9

Source: Adapted from Australian Petroleum Production & Exploration Association Limited (2020: 4).

TABLE 2.4 Production of Natural Gas by State/Territory (Billion Cubic Feet), 2019

State/Territory	Conventional gas	Coal seam gas	Liquid natural gas exports
New South Wales	–	4.5	–
Northern Territory	450.7	–	393.7
Queensland	3.8	1,397.4	1,169.3
South Australia	97.7	–	–
Tasmania	0.7	–	–
Victoria	314.0	–	–
Western Australia	2,719.7	–	2,327.4
Total	3,596.6	1,401.9	3,890.4

Source: Adapted from Australian Petroleum Production & Exploration Association Limited (2020: 5).

prices went from $4 per gigajoule up to a peak of $20 per gigajoule in 2017, and then back to $10 per gigajoule in 2019.

(Reucassel 2020: 78)

In terms of natural gas, Australia has undergone a recent large increase in LNG exports.

In addition to coal, Australia has enormous reserves of coal seam gas (CSG) reserves scattered about the country. Coal seam gas (CSG) refers to methane that is trapped within pores and fractures in underground coal deposits. It is extracted by drilling a well into a coal seam, fracturing the seam hydraulically, releasing the gas by reducing the water pressure by pumping out the water, in a process gener-ally referred to as "fracking." Coal seam gas can be used in the same manner as natural gas, for heating and cooking. Conversely, coal seam gas is often confused with shale gas, the latter being methane held within shale layers rather in a coal seam. In contrast to the United States, shale gas production is nonexistent in Aus-tralia, despite the fact that there are large shale gas reserves in South Australia, the Northern Territory, and Western Australia. Coal seam gas makes up for 27 percent of Australian gas reserves and is an industry that appears to be in ascendancy and supported by both Coalition and ALP governments at the federal and state levels. Conoco conducted the first stand-alone commercial production of CSG in 1996 at a site adjacent to the Moura coal in Queensland. CSG exploration continues in Queensland, particularly in the Bowen, Galilee, and Surat Basins, as well as the Sydney, Gunndah, Gloucester, and Clarence-Moreton Basins of New South Wales. CSG is also being explored in South Australia, Tasmania, Victoria, and Western Australia. The production, supply, and use of coal seam gas results in greenhouse gas emissions. This includes methane released from exploration drilling, produc-tion testing and well completion, and gas production activities including process-ing, venting, and flaring.

The ecological and emissions footprints of Australia

Despite its relatively small population, Australia has made a disproportional contri-bution to climate change because its ecological and carbon footprints are so large. Thus, despite its relatively small population, Australia is part and parcel of what John Bellamy Foster aptly terms *catastrophe capitalism*. He asserts:

> Since the late 1980s. the world has been engulfed in an epoch of catas-trophe capitalism, defined as the accumulation of imminent catastrophe on every side due to the unintended consequences of "the juggernaut of capital. Catastrophe capitalism in this sense is manifested today in the convergence of (1) the planetary ecological crisis, (2) the global epidemiological crisis, and (3) the unending world economic crisis. Added to this are the main features of today's "empire of chaos," including the extreme system of impe-rialist exploitation unleashed by global commodity chains; the demise of the

relatively stable liberal-democratic state with the rise of neoliberalism and neofascism; and the emergence of a new age of global hegemonic accompanied by increased dangers of unlimited war.

(Foster 2020b: 1)

As a minor core country or a semi-peripheral country situated in the capitalist world system, Australia finds itself in a political-economic juggernaut situated between two superpowers embedded in a new cold war, namely its erstwhile military ally the United States and China, the factory of the world, to which it sells iron ore and coal in huge amounts along with wheat and meat products and from which it buys numerous consumer products and imports university students in order to prop up its higher education sector (Smith 2020).

The term "ecological footprint" refers to the human impact upon the environment based on productive land and marine required to support the overall material standard of living of a particular region or country. It is calculated by measuring the consumption of energy, biomass (such as food and fiber), building material, water, and other resources and converting them into land area. Table 2.5 depicts the ten countries with the highest footprints, with Australia ranking number nine.

Table 2.6 depicts the 20 leading countries in carbon dioxide emissions from fuel combustion. While Australia ranks 15th in this list, bear in mind that all of the countries in the list have populations that are considerably larger than that of Australia.

Table 2.7 depicts Australia's greenhouse gases for selected years between 2000 and 2020, by financial year.

Table 2.8 depicts Australia's direct greenhouse gas emissions by economic sector in 2018.

According to the federal government's statistics between 1990 and 2018, direct emissions between 1990 and 2018 from mining increased 112.6 percent; direct emissions from services, construction, and transportation increased 53.4 percent; and direct emissions from the residential sector increased 40.9 percent, and

TABLE 2.5 Countries with the Highest Ecological Footprints

Rank	Country	Ecological footprint in global hectares per capita
1	United Arab Emirates	10.68
2	Qatar	10.51
3	Bahrain	10.04
4	Denmark	8.26
5	Belgium	8.00
6	United States	8.00
7	Estonia	7.88
8	Canada	7.01
9	Australia	6.84
10	Iceland	6.50

Source: Pariona (2017).

TABLE 2.6 Leading 20 Countries in Carbon Dioxide Emissions from Fuel Combustion

Rank	Country	Carbon dioxide emissions (in million tons)
1	China	9,300
2	United States	4,800
3	India	2,200
4	Russia	1,500
5	Japan	1,100
6	Germany	718.8
7	South Korea	600
8	Iran	567.1
9	Canada	548.8
10	Saudi Arabia	532.2
11	Indonesia	496.4
12	Mexico	446.0
13	Brazil	427.6
14	South Africa	421.7
15	Australia	384.6
16	Turkey	378.6
17	United Kingdom	358.7
18	Italy	321.5
19	France	306.1
20	Poland	305.8

Source: Adapted from World Population Review (2021).

TABLE 2.7 National Inventory Total from 2000 to 2020, by Financial Year

Financial Year	Emissions (Mt CO_2e)
2000	555.2
2003	594.8
2005	630.1
2007	651.1
2010	606.8
2013	556.4
2016	536.7
2018	543.1
2019	534.4

Source: Adapted from Australian Government, Department of Industry, Science, Energy and Resources (2021: 60).

TABLE 2.8 Australia's Direct Greenhouse Gas Emissions by Economic Sector, 2018

Sector	Emissions (Mt CO_2e)	Share of total emissions (%)
Primary industries	156.4	29.1
Agriculture, forestry & fishing	106.2	19.8

(Continued)

TABLE 2.8 (Continued)

Sector	Emissions (Mt CO₂e)	Share of total emissions (%)
Agriculture, forestry, & fishing changes in inventories in forest & wood product stocks	–44.7	–8.3
Mining	94.9	17.7
Manufacturing	59.4	11.1
Electricity, gas, water & waste services	189.8	35.3
Services, construction & transportation	62.2	11.6
Residential	69.6	13.0
All sectors	537.4	100.0

Source: Australian Government – Department of Industry, Science, Energy and Resources (2020: 4)

direct emissions from electricity, gas, and waste increased 28.2 percent (Australian Government – Department of Industry, Science, Energy and Resources 2020: 4–5). Conversely, emissions between 1990 and 2018 from agriculture, forestry, and fishing declined 62.9 percent and emissions during the same time period from manufacturing declined 17.7 percent (Australian Government – Department of Industry, Science, Energy and Resources 2020: 4–5).

Energy – electricity

Most electricity in Australia is generated at coal-fired power plants. However, Tasmania derives its electricity from hydroelectric generation in which rainwater dams and lakes store huge amounts of potential energy, which is periodically released through electricity-generating turbines with the water flowing downstream to the sea. Hydro Tasmania manages all hydroelectric generation in Tasmania cooperatively. At their peak, coal-fired plants provided 83 percent of electricity in Australia (Warren 2019: 6).

Coal-fired plants which rely on either black and brown coal reportedly supply about 60 percent of the total electricity production in Australia, but the reliance on such plants varies considerably among the various political jurisdictions. Most of Australia's coal-fired power plants were constructed in the 1970s and 1980s by state-owned electricity bodies. State electricity commissions were created in Victoria in 1924, in Queensland in 1938, in Western Australia in 1945, in South Australia in 1945, and in New South Wales in 1950 (Warren 2019: 34). South Australia constructed a coal-fired plant at Port Augusta, near brown-coal deposits at Leigh Creek. Victoria increased the number of coal-fired power plants in the Latrobe Valley to three. New South Wales built coal-fired plants in the Hunter Valley, a region rich with black coal deposits. Queensland constructed them in Darling Downs and west of Rockhampton, and Western Australia near the Collie coal deposits.

Beginning in the 1990s, the adoption of neoliberalism by state governments resulted in the privatization of these facilities and incorporated into the new National Energy Market by which generators compete against each other. The Coalition Kennett government sold Victoria's generators and network assets in 1995, using the income to pay off debts and build public infrastructure (Warren 2019: 40). The South Australian government sold its electricity assets in 1999 to pay off debts. Private energy companies in Australia include AGL and Origin (both Australian-owned), TXU (US-owned), International Power (British-owned), China Light and Power (Hong Kong-owned), and Singapore Power.

Table 2.9 depicts the assemblage of existing coal-fired power plants in Australia.

The National Energy Market (NEM) serves about 21 million out of 25.7 million Australians and covers the eastern seaboard from Ceduna in South Australia to Port Douglas in northeast Queensland and extends through a series of undersea cables and overland lines to Bruny Island of the eastern coast of Tasmania to Broken Hill in western New South Wales, and up to Caraldine in Central Queensland (Warren 2019: 40). Essentially, the Australian Energy Market Operator (AEMO) administers the NEM as a "collection of poles and wires, power stations and power

TABLE 2.9 Coal-Fired Power Plants in Australia

	Year Commissioned	*Year Scheduled to Close*
New South		
Bayswater	1982	2035
Eraring	1982	2034
Liddell	1971	2022
Mt. Piper	1993	2043
Vales	1978	2028
Queensland		
Callide B	1989	2039
Callide C	2001	–
Gladstone	1976	2032
Koran Creek	2007	2042
Millmerran	2002	–
Stanwell	1993	2046
Tarong	1984	2036
Tarong B	2002	
Victoria		
Loy Yang A	1984	2048
Loy Yang B	1993	2046
Yallourn	1975	2032
Western Australia		
Collie	1999	2040
Muja	1981	2022 for units 1–5
		2024 for unit 6
		2040 for units 7–8
Blue Waters	2009	

Source: Australian Government (2018).

point, turbines and transformers variously owned by governments, more than 200 companies and by every home and building owner in the grid" (Warren 2019: 91). Electricity generators, ranging from coal-fired power plants to wind farms to roof-top solar, bid into the market. Electricity demand is highly weather-dependent, increasing on very hot or very cold days and declining in spring and autumn. The small Newport Power station in Melbourne functions as an intermediate gas generator.

The South Australian electricity grid is the smallest of the mainland grids (Warren 2019: 78). It is connected to the Victorian grid by two transmission lines or interconnectors. South Australia has three main power stations, namely the Northern Power Station at Port Augusta and two large gas generators at Torrens Island and Pelican Island. Small grids also include the North-West Interconnected System in the Pilbara region of Western Australia, the Darwin-Katherine grid in the Northern Territory, ones in the Perth area, Mt. Isa in Queensland, Alice Springs in the Northern Territory, Broome in Western Australia, Coober Pedy in South Australia, King Island in the Bass Strait and hundreds of tiny grids in remote communities and cattle stations (Warren 2019: 183).

Over the past decade or so, due to public pressure, including the climate movement, federal and state government subsidies, and falling prices, renewable energy has increased sharply since 2012, but accounts for only 19.8 percent of Australia's electricity generation (Monash Sustainable Development Institute 2020: 7). A more profound increase has occurred in South Australia. Warren reports:

> South Australia was per capita the biggest rooftop solar PV state in the country, ahead of Queensland and Western Australia. By 2016, solar was on more than 30 percent of all dwellings in South Australia and Queensland. In some Adelaide suburbs up to 70 percent of households have solar PV. Household solar systems were much larger in number but smaller in total generation than wind farms. By 2016, around seven per cent of South Australia's total electricity supply came from solar PV, while 34 per cent came from wind.
>
> *(Warren 2019: 76)*

Whereas only about 8,000 Australian households had rooftop solar photovoltaic panels in 2007, this figure has skyrocketed to over 2 million households, accounting for 20 percent of the Australian residential market and around 70 percent in Brisbane and Adelaide (Warren 2019: 183). For a number of reasons, since 2012, ten coal-fired power plants have been closed in Australia, and it is anticipated that half of the remaining ones will be closed because they would have reached their use-by dates (Warren 2019: 14). The Northern coal-fired plant at Port Augusta closed in part due to its difficulty competing in the new renewable energy market and public pressure, including from the climate movement, in South Australia (Warren 2019: 99).

Energy – stationary energy excluding electricity production

Stationary energy excluding electricity includes emissions from direct combustions of fuels, predominant from manufacturing, mining, residential, and commercial sectors. This includes emissions from houses, apartments, offices, schools, universities, hospitals, hotels, residences, shops, restaurants, cafes, cinemas, churches and other religious structures, etc. Indirectly most of these emissions are a by-product of electricity draw from coal-fired power plants for heating during cold weather and cooling during hot weather. Green and Minchin (2010: 130) report:

> [T]he impact of buildings is usually underestimated in the climate debate, largely because greenhouse gas emissions from buildings are generated indirectly, from the materials used to construct them to their use of electricity from coal-fired power stations. Even something as simple and commonplace as cement has a huge impact: the cement industry along accounts for about 4 percent of global greenhouse gas emissions. More than anything, buildings are surprisingly large greenhouse contributors because of the massive and often very wasteful use of energy inside of them, particularly the use of inefficient heating, cooling, ventilation and lighting.
>
> *(Green and Minchin 2010: 130–132)*

Many modern buildings do not allow for ventilation in that windows cannot be opened allowing fresh air into them, which prevents regulating air flow depending on outside temperatures, resulting in excessive heating and air conditioning depending on the season, and a major health problem during the COVID-19 pandemic.

Energy – transportation

Transportation is reportedly the second-largest source of greenhouse gas emissions in Australia, next to electricity generation (Stock et al. 2018). Emissions from road, rail, domestic aviation, and domestic shipping accounted for 17.6 percent of emissions in year to September 2020. They decreased over the course of the year due to less driving as a result of the COVID-19 pandemic. Eighty-five percent of transport emissions come from road transport and over 60 percent from automobiles and light commercial vehicles. In 2018, the transport emissions for the various Australian political jurisdictions were:

- Western Australia – 14.5 million tons or 5.6 tons of CO_2/person
- South Australia – 6.6 million tons or 3.8 tons CO_2/person
- Victoria – 22.3 million tons or 3.6 tons CO_2/person
- Northern Territory – 1.3 million tons or 5.3 tons CO_2/person

- Tasmania – 1.7 million tons or 3.3 tons CO_2/person
- Queensland – 22.5 million tons or 4.6 tons CO_2/person
- New South Wales – 27.4 million tons or 3.51 tons CO_2/person
- ACT – 1.2 million tons or 3.01 tons CO_2/person (Stock et al. 2018: 5).

In comparison to most other advanced capitalist countries, Australia's motor vehicle performance ranks poorly due to a lack of greenhouse gas emissions standards (or fuel-efficiency standards) in place, relatively high distances traveled per person by car, a low share of trips taken on public, and low government spending on public transport. As a large country with far-flung populations and a substandard inter-city railway system, Australians rely heavily upon air travel, not only domestically but internationally. The Department of Infrastructure and Regional Development reports:

> In the 2015–16 period, more than 36.2 million people and 996,000 tonnes of freight were carried on international air services to and from Australia. In the same period 60.9 million passengers were carried on domestic flights. Based on 2010–11 data, the Australian Government of Infrastructure, Transport and Regional Economics (BITRE) has projected the number of passenger movements from all Australian airports to increase by 3.7 percent per annum over the next 20 years, from 135.1 million in 2010–11 to 279.2 million in 2030–31. International and domestic passenger movements are projected to increase by 4.9 and 3.3 per cent respectively per annum over the same period to 72.1 and 207.1 million, respectively in 2030–31.
>
> *(Department of Infrastructure and Regional Development 2017: 2)*

In 2016, Australia's domestic aviation sector produced 9.99 Mt of CO_2e and Australia's international aviation sector produced 12.02 Mt of CO_2e (Department of Infrastructure and Regional Development 2017: 11). Between 2012 and 2016, the passenger revenue-ton-kilometers (RTK) increased 5.1 percent and the cargo revenue-ton-kilometers (RTK) increased 4.2 percent, indicating a growing reliance on both on aviation (Department of Infrastructure and Regional Development 2017: 9). However, the COVID-19 pandemic in 2020 greatly curtailed international flights in and out of Australia as well as inter-state flights due to state border lockdowns that were gradually lifted in late 2020 as the severity of the pandemic subsidized in Australia, in contrast to many other countries elsewhere in the world. The airline industry and both the federal and state governments are bucking to return to business-as-usual, including in rejuvenating the lucrative overseas student market in Australian universities.

Energy – fugitive emissions

Fugitive emissions occur during the production, processing, transport, storage, transmission, and distribution of fossil fuels, including coal, crude oil, natural gas, and decommissioned underground coal mines. Sources of fugitive emissions include CO_2 co-mingled naturally with methane and hydrogen in natural gas and released in gas processing, CO_2 released from natural gas combustion in gas processing, and methane released from coal mining (Garnaut 2019: 127). This sector also includes emissions from decommissioned underground coal mines.

Industrial processes and product use

Emissions from industrial processes and product use are the by-products of material and reactions using in production processes, including chemical, metal, and mineral products, and emissions from consumption of synthetic gases. For example, the conversion of bauxite into aluminum requires a lot of electricity. Warren reports:

> At its peak, the aluminium sector consumed 14 per cent of Australia's total electricity. Their sheer buying power enabled them to cut deals at extraordinarily low prices.
>
> *(Warren 2019: 38)*

In contrast to other countries where aluminum smelting relies on hydroelectric power, the Australian aluminum industry has the highest greenhouse gas intensity in the world, at its peak accounting for around 6 percent of the country's total emissions (Davison 2004: 125). Australia produces about 2.5 percent of total global aluminum production (Garnaut 2019: 114). However, it is the world's largest producer of alumina, about a quarter of the world's total, most of which is exported to overseas smelters. Australian alumina mining occurs in two locations, namely southwestern Western Australia and the Gladstone in Central Australia, where the bauxite from the Gulf of Carpentaria is mined. Aluminum smelting in Australia relies upon electricity which is derived from coal-fired power plants.

Agriculture

Despite the fact that most of Australia is arid or semi-arid, the country's agriculture is highly diverse and in large part energy-intensive and heavily dependent upon scarce water resources, depending upon the sector, whether it involves the production of grain crops, sheep and wool, beef and diary items, working bullocks and horses, sugar cane, cotton, fruit and vegetable, and grapes and wine (Henzel 2007). Agricultural emissions include methane, nitrous oxide, and carbon dioxide. Methane and nitrous oxide emissions result from enteric fermentation and management of livestock and also are the by-products of rice cultivation, agricultural soils, and the burning of agricultural residues. CO_2 is a by-product of the application of urea and lime on the soil.

Approximately 80 percent of Australia's agricultural emissions come from cattle and sheep production (Reucassel 2020: 182). In 2019, Australia had 24.7 million head of cattle and a sheep flock of 65.76 million head (Meat & Livestock Australia

2020: 17, 21). In terms of domestic consumption, one kilogram of beef produces around 23 kilograms of CO_2 and one kilogram of lamb produces slightly less than 10 kilograms of CO_2 (Reucassel 2020: 181). Digestive bacteria in the stomachs of both cattle and sheep release methane. Reportedly, a "kilogram of chicken creates around 16.9 per cent of the greenhouse gases as the same amount of beef" (Reucassel 2020: 185). In 2019, Australia was the second-largest beef exporter, after Brazil, and the world's largest sheep meat exporter (Meat & Livestock Australia 2020). Furthermore, in 2019 it exported over 1.3 million live cattle and 1.1 million live sheep. In 2019, 27 percent of Australia's beef exports went to China, 23 percent to Japan, and 22 percent to the United States (Meat & Livestock Australia 2020: 19). Dairy products vary in terms of their carbon footprint. Whereas a liter of milk results in over one kilogram of CO_2, cheese results in over eight kilograms of CO_2 (Reucassel 2020: 189).

Finally, quite a bit of fresh fish is flown around Australia and in and out of the country, thus contributing to greenhouse gas emissions due to air transportation. Reucassel (2020: 188) reports: "Fish can definitely be at the lower end of the scale for meat and greenhouse gases, but proper labelling will be essential to make sure you aren't getting a deep-sea traveled, frozen and airfreighted fillet that could be as bad for your carbon footprint as your bite of beef."

Waste

Emissions from waste include emissions from landfills, wastewater treatment, waste incineration, and biological treatment of solid waste – largely methane, which is generated when organic matter decays under anaerobic conditions. Reuscassel (2020: 204) reports: "The capture of methane from landfill to create electricity was encouraged by the Emissions Reduction Fund," a component of the federal government's Direct Action Plan. The larger challenges are minimizing food waste and composting it which will release biogenic CO_2, but this is minor compared to methane release at a land fill.

Land Use, Land Use Change and Forestry (LULUCF)

The LULUCF sector consists of emissions from forests and agricultural lands and changes in land use. The Australian government reports:

> Net emissions for the LULUCF sector in the year to September 2020 are estimated to be -18.8 Mt CO_2e. This net sink has declined by 4.1 per cent (0.8 Mt CO_2e) on the previous twelve months due to continuing declines in land clearing and increase in emissions from agricultural soil.
>
> *(Australian Government, Department of Industry, Science, Energy and Resources 2021: 19)*

Drought conditions for several years over much of Australia also contributed to a decline of land use, at least until the beginning of 2020. Mackey et al. (2008), a team of ANU scientists, argue that the National Carbon Accounting System underestimates the carbon-carrying capacity of natural forests with high biomass stocks.

Greenhouse gas emissions that Australian governments do not directly account for

In this section, I discuss three sources of greenhouse gas emissions that Australian governments do not factor into their reporting and that are seldom discussed in the media or the general public. They are emissions resulting from fossil fuel exports, tourism, and international education.

Emissions from fossil fuels exports

Various studies have indicated that Australia outsources more CO_2 emissions from coal exports than it generates domestically. Guy Pearse states:

> Our [coal] exports alone generate almost 700 million tonnes of CO_2 annually, which is almost 15 percent more than Australia's entire annual greenhouse-gas emissions. Yet currently our coal exports are projected to nearly double [by] 2029–30.
>
> *(Pearse 2009: 75)*

More recently, Swann (2019: 20) reported that "Australia exports fossil fuels equal to 1.1 billion tons of CO2 per year," more than double its domestic emissions. The RMIT ABC Fact Check Unit (2019) estimated that Australia's domestic emissions plus the emissions embedded in its exports added to 1,712 million tons in 2016 or approximately 3.6 percent of total global greenhouse gas emissions for that year, considerably more than the 1.2–1.3 percent figure based solely on domestic emissions figure generally cited by government agencies and politicians for Australia's contribution to global greenhouse gas emissions. However, when Fact Check removed Australia's imported fossil fuel emissions, it reduced Australia's contribution to total emissions to 3.3 percent. No matter how the emissions are counted, Australia is punching above its weight to greenhouse gas emissions and anthropogenic climate change.

Tourism

Tourism is a double-edged sword, being a victim of climate change and a contributor to it.

Since the advent of cheap air flights, tourism has skyrocketed, becoming an integral component of the culture of consumption. Many countries and regions, such as the Caribbean, Europe, Latin America, Southeast Asia, Australia, New Zealand, the Pacific, rely heavily upon tourism as a source of income. While some tourists, such as backpackers, seek to travel simply and by train, coach, or ferry, many rent cars, or fly hither and thither on airplanes, or take luxury cruise ships and stay at luxury resort hotels that require tremendous energy and resources to operate, resulting in greenhouse gas emissions and contributing to climate change. The United Nations World Tourism Organization and the International Transport Forum report:

> In 2016 . . . international tourist arrivals reached 1.2 billion, representing a 65% increase from 2005. Domestic tourist arrivals reached 8.8 billion, representing a 119% increase from 2005. In addition, international and domestic same-day visitors doubled in comparison to 2005 (counting then 10 billion). This means that in 2016, an estimated 20 billion tourist trips were taking place. This number translates into transport-related emissions from tourism of a total of 1,597 million tonnes of CO_2, with 1,371 million tonnes of CO_2 accounting for all overnight stays and same-day visitors accounting for 200 million tonnes. The total transport-related tourism emissions represented about 22% of the total transport emissions and 5% of the overall emissions (32,100 million tonnes) in 2016.
>
> *(United Nations World Tourism Organization and the International Transport Forum 2019: 43)*

Both domestic and international tourism are huge industries in Australia that generate largely unaccounted greenhouse gas emissions, particularly in the form of air travel within Australia and on the part of Australians taking holidays abroad and international tourists visiting Australia. Based upon 19 telephone interviews with representatives from federal government tourism departments, state tourism organizations, and federal and state tourism industry councils, Ruhanen and Shakeela maintain that the "Australian tourism industry is largely operating in 'business as usual' mode" in terms of its response to climate change (Ruhanen and Shakeela 2014: 46).

While most international tourists travel to and from Australia by air along with many domestic tourists traveling around the country by air, some international tourists travel to Australian ports by cruise ships. Australia is part and parcel of the international cruise ship market. Until the COVID-19 debacle involving the docking of the *Ruby Princess* in Sydney on March 19, 2020, cruise ships regularly embarked and disembarked from Australian ports, particularly Sydney, Brisbane, and Melbourne. McNab (2021: 2) reports: "The price of the unchecked spread of the virus among the *Ruby Princess* crew and passengers, was the lives of around twenty-two passengers and over 700 people infected." After disembarking the *Ruby Princess*, more of its passengers boarded air flights to Australian and overseas destinations without being properly tested for COVID-19.

International education

International higher education has become the third-largest export industry in Australia and Victoria's leading export industry, at least prior to the COVID-19 pandemic, only superseded by the export of iron ore and coal. Cutbacks in government funding since the 1980s have compelled Australian universities to derive an increasing amount of their incomes from full-fee paying overseas students, particularly from China, but also other developing countries, such as India, Indonesia, Malaysia, and Singapore. The growing internationalization of Australian higher education is highly dependent upon air travel, a growing source of greenhouse gas emissions, thus belying frequent claims on the part of Australian universities to being committed to environmental sustainability. While in 2015, Australia's total investment, both public and private, in tertiary education as a share of GDP was "above the OECD average and the fourth highest, behind the US, Chile and Canada," its public investment in higher education as a "share of GDP was among the lowest in the OECD – ranked 24 out of 34 countries" and was "0.77 per cent of GDP in 2015 compared to 0.98 per cent of GDP for average OECD countries" (Universities Australia 2019: 14).

The Australian Bureau of Statistics indicates that "in 2017–18, international education was worth $32.4 billion to the Australian economy, up from $28.1 billion in 2016–17" (Parliamentary Library 2019: 1). Furthermore, the Parliamentary Library reports:

> Higher education accounted for 68.5 per cent ($22.2 billion) of international education export income in 2017–18 and 45.6 per cent of all overseas student enrolments in 2018. Overseas students also enrol in Australian vocational education and training (VET), schools, English Language Intensive Courses for Overseas Students (ELICOS), and non-award courses. However, each of these accounts for few students and has less economic impact than overseas student enrolments in higher education.
>
> *(Parliamentary Library 2019: 1)*

During the period 2008–2018 alone, the number of overseas higher education students almost doubled from 202,581 to 398,553 (Parliamentary Library 2019: 2). Of these students in 2018, 38.3 percent (152,591) were from China, and 18.0 percent (71,857) were from India, with the remaining top ten nationalities being Nepal, Vietnam, Malaysia, Pakistan, Sri Lanka, Hong Kong, and Singapore (Parliamentary Library 2019: 3). In 2019, of 1,562,520 higher education students in Australia, 1,082,55 were domestic students and 479,987 were international students (Universities Australia 2020). One report indicated that several universities derived over 50 percent of their student fees revenue from international students in 2018, 58.5 percent in the case of the University of Sydney, 58.5 percent in the case of the University of Melbourne, 27.5 percent in the case of the University of New South Wales, 55.1 percent in the case of Monash University, 51.6 percent in the case the University of Queensland, and 51.2 percent in the case of Federation University of Australia – 51.2% (Carey, Prytz, and Hefferman 2020). International students' tuition fees reportedly provided 26 percent of all university revenue in 2018 in contrast to 10 percent in 2000 (Carey, Prytz, and Hefferman 2020).

While there is a growing concern about the contribution that academic air travel makes to greenhouse gas emissions, there has been relatively little research on the contribution that international and domestic students make in flying to universities or academic-related events away from home. As Shields observes,

> There are no data on the number of trips mobile students make to their home country. The number of trips could range from less than one (e.g., one trip to the home country every other year) to several trips per year.
>
> *(Shields 2019: 597)*

Obviously the number of trips that an overseas student can afford to take to his or her home country depends upon his or her finances, or at least those of his or her parents:

> The results indicate that GHG emissions associated with international student mobility were between 14.01 and 38.54 megatons of CO_2 equivalent in 2014, having increased from between 7.24 and 18.96 megatons in 1999. To contextualize these figures, the low estimate is comparable to the national annual emissions of Latvia (13.94 megatons) or Jamaica (15.47 megatons), while the upper estimate is comparable to those of Croatia (30.42 megatons) or Tunisia (39.72 megatons).
>
> *(Shields 2019: 598)*

Shields modeled two scenarios for international university student travel, a low one entailing one return trip per year and a high one entailing two return trips per year. The vast majority of these students make their way between their home countries and their host countries by airplanes, although it is difficult to say how often they make the trip, whether once or several times during the course of their overseas studies.

Conclusion

The National Inventory for annual greenhouse gas emissions in Australia has exhibited considerable fluctuation since 2000, increasing considerably during the Coalition government of John Howard years between 1996 and 2007, undergoing some decline during the ALP government years of Kevin Rudd and Julia Gillard, and more or less staying at the same level during the Coalition governments of Tony Abbott, Malcolm Turnbull, and Scott Morrison, until the economic downturn prompted by the COVID-19 pandemic. Compared to the high emissions peak of the Howard years, Australia has managed to reduce its overall annual emissions due to several factors: the phasing out of several coal-fired coal plants; the adoption of renewable energy, notably in terms of household and building photovoltaic solar panels, some solar power farms, and wind farms scattered around the country; and the brief operation of a Carbon Pricing Mechanism or modest tax in 2012–2014, the machinations of which will be discussed in detail in the next chapter. Due largely to the COVID-19 pandemic, Australia's greenhouse emissions declined for most sectors as is indicated in Table 2.10.

Unfortunately, Australian government statistics nor statistics from any other available source document the impact that social inequality on the emissions footprint, namely the fact that the wealthy and powerful emit more greenhouse gas emissions than poor people. However, on this score, I suspect that Australia is not greatly different from the United States. Chancel reports:

> In the United States, for example, average CO_2e emissions are twenty-three metric tons annually person: but the poorest 50 percent emit about thirteen metric tons of CO_2e per year and the wealthiest 1 percent emit as least 150 metric tons. This disparity is the result of a very energy-intensive model of consumption (even at the bottom of the social leader) with high levels of income and consumption dispersion.
>
> *(Chancel 2020: 96)*

During the height of the COVID-19 pandemic, most flights into Australia were grounded, and many workplaces, shops, restaurants, cinemas, sporting events, and schools and tertiary educational institutions went into lockdown. However, the containment of COVID-19 due to restrictions has resulted in a decline of COVID cases and deaths and an opening up of the Australian economy with a reopening of many of the activities that had been in total or partial lockdown. Many Australians are opting to drive rather than take public transit due to ongoing fears of contacting COVID-19 on trains, trams, and buses, resulting in a considerable increase in traffic, particularly in urban areas, and undoubtedly in an increase in transport emissions. While a shutdown of international flights other than returning

TABLE 2.10 The Impact of the COVID-19 Pandemic on Greenhouse Gas Emissions (Mt CO_2e) in Australia

Sector	Year to September 2019	Year to September 2020
Energy – Electricity	177.5	170.4
Energy – Stationary energy excluding electricity	100.4	101.8
Energy – Transport	100.1	89.8
Energy – Fugitive emissions	55.3	51.2
Industrial processes and product use	32.2	34.3
Agriculture	73.9	72.0
Waste	13.6	13.3
Land use, land change and forestry	-19.7	- 18.8
Total	533.4	510.1

Source: Adapted from Australian Government, Department of Industry, Science, Energy and Resources. 2021: 9.

Australians and some international athletes, such as was the case for the Australian Open in Melbourne in early 2021, domestic air travel has gradually been returning to the levels of pre-COVID-19 times as both the federal and state governments and the tourist industry seek to "snap back" in the words of Prime Minister Scott Morrison.

In Australia, there have been some economic activities in which the health standard has not applied as stringently as they have for small businesses, such as restaurants, cafes, pubs, and cinemas where people are more likely to gather in cramped quarters. These have included the mining industry, meat packing factories, the construction industry, professional sports (at least the players who, while not necessarily performing in packed stadiums and arenas, continued to perform for TV audiences), and a down-scaled-back airline industry. There is the danger the temporary blip in the global economy will prompt capitalists and their political allies worldwide as well as in Australia to attempt to ramp up the capitalist treadmill of production and consumption in the aftermath, whenever that might be, and further ignore the warnings of climate science.

3
CLIMATE POLITICS AT THE TOP

The corporations and the two
major political parties

Globally, Australia is distinct in that it has been the only country that adopted and then abandoned some form of carbon pricing. The dynamics of Australian climate politics have played out at various levels, at the top, more intermediate, and grass-roots levels. In this chapter, I explore Australian climate politics at the top, particularly in the relationships between the corporate sector, particularly the mining and energy production sectors, and the two major political parties, namely the Coalition Party and the Australian Labor Party (ALP).

Sometime ago Noam Chomsky said the United States has a "one party system," namely the Business Party which has two factions, the Republicans and the Democrats. In a similar vein, one might argue that Australia also has a Business Party of its own, with the Coalition and the Australian Labor Party (ALP) being its two major factions. In many ways, business and politics around the world are so interwoven that one might speak of national variants of state capitalism, US state capitalism, German state capitalism, Japanese state capitalism, Australian state capitalism, Chinese state capitalism, on and on. As Boswell and Chase-Dunn (2000: 86–87) assert, "all societies are 'state capitalist', if what is meant by this elusive term is using the state to promote economic development within the capitalist world-system."

While multinational or transnational corporations are at the center of the capitalist world system and their primary objective is profit-making, which they increasingly achieve through globalization of their operations, each multinational corporation retains close ties to its country of origin, which generally is a Western country, but may be a non-Western country, such as Japan, China, or India. While prior to perhaps the 1980s, the ALP with its strong alliance with the Australian labor movement constituted an oppositional force to the capitalist class, with the full-fledged acceptance of neoliberalism by the ALP Hawke and Keating governments between 1983 and 1996, the ALP entered into an unholy alliance with at least sections of the Australian capitalist or corporate class. Edwards (2020: xii)

DOI: 10.4324/9781003202530-4

asserts Australia is increasingly described as an oligopoly economy. Mega-corporations exert a profound influence on party politics at the both federal and state levels as well on the mass media and higher education.

Nevertheless, neither the Australian corporate class nor the two major political parties are monolithic entities. In terms of climate change, there are varying perspectives on climate change in both. While the mining and fossil fuels industries are filled with climate skeptics, the financial and insurance industries appear to have executives who accept the findings of climate science and assert some actions have to be taken on curtailing greenhouse gas emissions, albeit without seriously questioning the basic premises of global capitalism. Conversely, while ALP politicians overwhelmingly, if not unanimously, accept the findings of climate science, Coalition politicians range from outright climate skeptics to those who accept the findings of climate science, although with perhaps some qualifications.

Climate politics has adversely impacted every Australian prime minister, starting with John Howard. Both Coalition and ALP governments at both the federal and state levels have consistently subsidized the exploitation of Australia's coal, natural gas, and coal seam gas supplies and supported the expansion of coal exports. The fossil fuel and coal-based electricity generation industries have dominated "almost every greenhouse-related consultative committee by the federal government and its agencies" (Pearse 2009: 41). This was also the case for the federal ALP governments under Prime Ministers Kevin Rudd and Julia Gillard between 2007 and 2013, and in subsequent Coalition governments. Australian governments have traditionally subsidized the costs to private enterprise of large-scale development projects. Large state investments in electricity generation and ports and harbor developments often function as a subsidy to the private sector to encourage mineral exports, including coal. Mining companies have played a key role in Australian politics.

Over the years, the fossil fuels industry has systematically contributed to the major Australian political parties to curry favor to its economic interests. Table 3.1

TABLE 3.1 Fossil Fuel Companies' Funding for Major Political Parties in Australia, 2017–2018

Company	ALP	Liberal	National
Woodside	$117,700	$118,500	$1,100
Santos	$85,610	$71,975	$24,500
Chevron Australia	$59,000	$51,219	$11,660
Alinta Energy	$68,500	$25,000	$25,250
Origin Energy	$54,400	$39,670	$16,425
Caltex	$44,138	$52,769	$0
Minerals Council of Australia	$41,200	$43,250	$10,450
Cartwheel Resources	$0	$52,864	$0
Senex Energy	$36,500	$15,400	$0
St Baker Family Trust	$0	$50,000	$0

Source: Hinman (2019: 9).

depicts the funding provided by various fossil fuel companies for the Liberals, Nationals, and the ALP in the fiscal year 2017–2018.

During the 2018–2019 fiscal year, the fossil fuel industry donated $85,719,747 to Australian political parties, with $83.7 million of this amount being a donation from Clive Palmer's Minerology to the mining magnate's own United Australian Party (UAP) (Australian Conservation Foundation 2020: 6). Of the remaining roughly $1.9 million, $1,147,376 went to the Coalition and $725,448 to the ALP.

Like in other developed capitalist countries, particularly the United States, in Australia corporate-linked lobbyists exert a strong influence on politicians, particularly ones belonging to the two major parties. For the period 2007–2016, resource sector lobby group revenue came to $203,594,120 for the Minerals Council of Australia (MCA), $138,594,120 for the Australian Petroleum Production and Exploration Association (APPEA), $84,804,591 for the Queensland Resources Council, $76,936,909 for the New South Wales Minerals Council, and $36,944,082 for the Australian Coal Association (ACA), which was absorbed by the MCA in 2013 (Aulby 2017: 9). Furthermore, 10 of 14 board positions of both the Minerals Council of Australia and the Queensland Resources Council are from foreign-owned corporations, including Peabody, Anglo American, BHP, Rio Tinto, Glencore, and Adani Mining, yet another indicator of the influence of overseas-based multinational corporations on Australian politics (Aulby 2017: 11).

The ALP government and the coalition opposition, 1983–1996

According to Sklair (2002: 5), during the "1980s and 1990s Australia made a rapid and deep-reaching transition from a high tariff, ultra-protectionist inward-looking economy and society to one of the most open economies in the world." Ironically, much of this transition occurred under the Hawke and Keating governments between 1983 and 1996. Since the late 1970s, Coalition and ALP governments at federal and state levels increased support for the exploitation of Australia's coal and natural gas supplies (Corrighan 1980; Pearse 2009). In 1984 Australia overtook the United States as the world's largest exporter of coal: a dubious status that it retained until 2011 when Indonesia, with Australia slipping to the number two position (Department of Resources and Energy 1986: 5; World Coal Association 2015). When the Department of Resources and Energy (1986) acknowledged community concerns about the detrimental environmental consequences of coal mining and utilization, it recommended the government provide advice and assistance to the coal industry in the form of research and development on environmental matters.

In 1987, the labor movement engaged in numerous strikes and 24-hour work stoppages in support of the creation of a National Coal Authority, which would have "centralised planning of investment, mine development, and productive capacity, backed by the ability to directly control mining operations," but the ALP government ignored this demand (Lee and Draper 1988: 49). According to Lee and Draper,

> [T]he Hawke government has consistently espoused the philosophy of the free market, even at the expense of the ALP Platform. Insofar as any attention is paid to planning it is by way of powerless industry consultative councils or plans to deregulate industries. In the coal industry this has been manifest in the unwillingness of the Hawke government to honour the ALP policy to set up a National Coal Authority, and its reliance instead on the Australian Coal Consultative Council.
>
> *(Lee and Draper 1988: 47)*

Although export controls had been implemented in 1978 "in the wake of price cutting by Utah and in response to action . . . by mining unions," in September 1986 the Hawke government again worked against unions and curtailed the regulation of coal exports (Lee and Draper 1988: 48). State railway systems in Queensland and New South Wales served as carriers of coal from mines to various destinations, including coal ports. Queensland Rail was created in 1995 and FreightCorp, which was created in 1996 as a New South Wales state-owned freight operator, transports coal to the ports of Newcastle and Port Kembla, additional examples of state support of the coal industry (Energy Minerals Branch 1999: 7–9).

Ironically, during a period of growing neoliberalism or economic rationalism in Australia, there was growing awareness and concern about the "greenhouse effect" among many Australians. In 1980 the Australian Academy of Science convened a conference in Canberra to examine the growing sciences on the greenhouse effect (Pearman 1980). In 1986 federal science minister Barry Jones (1983–1990) established the Commission for the Future, which considered the greenhouse effect among other issues (Taylor 2014: 20). Based upon their reportage of a two-year media and public awareness campaign, Henderson-Sellers and Blong (1989: 155) reported that "the awareness of the greenhouse issue is probably greater amongst the general public in Australia than in any other country in the world." In addition to their book on the greenhouse effect, two other books appeared in 1989 penned by Australians, namely *The Greenhouse Challenge* by Jim Falk and Andrew Brownlow and *Living in the Greenhouse* by Ian Lowe (Falk and Brownlow 1989; Lowe 1989). Despite his commitment to economic growth and neoliberalism, ALP prime minister Bob Hawke stated in his Statement of the Environment in 1989:

> The growing consensus amongst scientists is that there is a strong possibility of global warming with major climate change, and that this is linked with the levels and nature of industrial and agricultural activity. Significant climate change . . . would have major ramifications for human survival.
>
> *(Hawke 1989)*

The Hawke government created the National Greenhouse Advisory Committee in April 1989 as a component of its National Climate Change Program (Taylor 2014: 47). In October 1990 Australia adopted the targets of the Toronto conference by creating an interim planning target of stabilizing greenhouse at "1988 levels by 2000, and reducing them by 20 per cent from that level by 2005" (Taylor 2014: 31).

The 1990 federal election exhibited a concern about climate change mitigation targets on both sides of the political divide (Taylor 2014: 2). While Graham Richardson, the environment minister, "had convinced Cabinet prior to the election to adopt the Toronto position as an 'interim' target," Andrew Peacock, the Liberal leader, "took an even more ambitious position to the 1990 election: that emissions levels would be cut by 20 per cent by 2000, rather than 2005" (Butler 2017: 23). However, these proposals became sidelined as neoliberalism with its focus on privatization and deregulation became ramped up even further during the Keating government (1991–1996) years. Cleary (1991) maintains that Keating served as "one of the prime movers in deferring a final decision on the now-famous Toronto target, a 20 per cent cut in emissions from 1988 levels."

The Coalition government and the ALP opposition, 1996–2007

As one of Australia's longest-serving prime ministers, John Howard adopted by and large a climate change denial stance, asserting that the reality of climate change remained open to scientific argument. In the lead-up to the UN Framework Convention on Climate Change (FCCC) conference in Kyoto in 1997,

> Howard instructed Australia's Environment Minister, Robert Hill, to negotiate a position that allowed Australia to increase its pollution levels, and to count reductions in broadscale land clearing that were already underway. Remarkably, Hill was able to achieve both outcomes – making Australia the only nation permitted to increase its emissions during the period (by 8 per cent, compared to an average reduction across the developed world of about 5 percent) and to count the benefit of land clearing reductions through a provision known as "the Australia clause".
>
> *(Butler 2017: 24–25)*

In contrast, under the provisions of the Kyoto Protocol, the European Union agreed to cut their emissions 8 percent from 1990 levels by 2012, Japan by 6 percent, and the United States by 7 percent, the latter largely due to the efforts of Vice President Al Gore.

Senator Robert Hill and his assistant Roger Beale proposed Australia's first Renewal Energy Target, calling for a modest additional two percent of energy from renewable sources by 2010 as well as a price on "carbon pollution," the latter of which John Howard rejected (Wilkinson 2020: 14). However, lobbyists representing the fossil fuels industries in both Washington, DC, and Canberra pushed strongly against the Kyoto Protocol. Thus, along with the United States with George W. Bush as president, Australia became one of the only two developed countries that did not ratify the Kyoto Protocol, although both countries had been signatories to it. Like Bush, Howard favored purported technological solutions, such as "clean coal" and carbon capture and sequestration rather than carbon pricing as strategies for reducing greenhouse gas emissions. In keeping with the state's support of the coal mining industry, the Howard government claimed it was collaborating with the coal industry in adopting "no-regrets measures to reduce emissions" generated by the burning of coal, and arguing that "uncertainty remains about the potential for emissions to contribute to a global warming effect" (Energy Minerals Branch 1999: 14). The government asserted that "[i]ncreasing combustion efficiency is an effective means of reducing emissions," and this could be achieved by "using coals with high caloric values," which were supposedly available in Australia (Energy Minerals Branch 1999: 15). To its credit, the Howard government established in 1998 the Australian Greenhouse Office, which was the world's first government agency committed to reducing greenhouse gas emissions.

During the Howard government years, a consortium of industries involved in the production of fossil fuels and coal-based electricity generation formed the self-designated "Green House Mafia," which "dominated almost every greenhouse-related consultative committee established by the federal government and its agencies" and hence dominates climate policy (Pearse 2009: 41). Members of the Greenhouse Mafia included the Australian Coal Association (ACA), the Minerals Council of Australia (MCA), the Business Council of Australia (BCA), and the Australian Industry Greenhouse Network (AIGN). The ACA represents black-coal production in New South Wales and Queensland and by the 1990s had become a major donor to both the Liberal and ALP parties. The AIGN reads like a *Who's Who* of the Australian corporate sector, particularly the fossil fuels industry, and includes or has included the Australian Aluminum Council, the Australian Coal Association, the Australian Petroleum and Exploration Association, the Minerals Council of Australia, the National Generators Forum, Alcoa World Aluminum-Australia, BP Australia, Caltex, Chevron Australia, ExxonMobil, Origin Energy, Rio Tinto Australia, Santos, Shell Australia, Woodside Petroleum, and Xstrata Coal Australia. Many AIGN lobbyists and climate skeptics operated under the Hawke and Keating governments and assisted in formulating Australia's coal export policies and continued to operate under the Howard government (Pearse 2009: 44).

In 2003 the Australian Coal Association formed the COAL21 Fund, a consortium of representatives from the coal and electricity industries, unions, federal and state governments, and the research community. The COAL21 Fund planned to raise over $1 billion over a ten-year period from a voluntary levy on coal production to support the pre-commercial demonstration of low emissions technologies in the power generation sector and research and development.

The Australian Bureau of Agricultural and Resource Economics (ABARE) has, since the mid-1990s, been the principal government body providing information on Australia's greenhouse gas emissions. Pearse (2009: 31) observes that the agency is advised by

> a who's who of fossil-fuel producers, burners and users [who] bought chairs on an ABARE steering committee (literally bought: the price was $500,000 per year, and payers include the Australian Coal Association, the Australian Aluminum Council, BHP, CRA [Cozinc Riotinto of Australia Ltd.], the Business Council of Australia, the Electricity Supply Association of Australia, Exxon Corporation, Mobile Australia and Texaco.

He goes on to note that "[w]hen carbon-lobby recruits are not moving through the revolving door between government and industry, they've often been moving sideways between industry associations in a game of musical chairs" (Pearse 2009: 40–41).

During the Howard government years, ALP state governments supported energy efficiency and renewable energy programs. Butler (2017: 25) observes:

> They also established the National Emissions Trading Taskforce in 2004 to build pressure for a market mechanism to bring down carbon pollution, and the ALP Carr government in New South Wales enacted one of the world's first carbon trading schemes through the GGAS program.

Howard became an energy booster, describing in 2006 "Australia as a potential 'energy superpower': the biggest exporter of coal in the world, soon to be the biggest exporter of gas, and with abundant wind and sunshine to power renewable technologies" (Warren 2019: 3). By 2006, spurred by rising global temperatures and growing public consciousness about the seriousness of climate change, as the 2007 federal election approached, he relented to public pressure in Australia and called for the creation of an emissions trading scheme to take effect in 2011. Howard created a task force to investigate emissions trading and announced in June 2007 that if he won the federal election he would implement an emissions trading scheme (ETS). In the interim, in January 2007, Howard appointed Malcolm Turnbull as minister of the environment and water. Malcolm Turnbull, an MP from the affluent seat of Wentworth in Sydney and a newcomer to party politics, accepted climate science and was an advocate of an ETS. He had cut his teeth in journalism, law, finance, and as the leader of Republican movement, an aborted effort to transform Australia from a constitutional monarchy to a republic.

Turnbull urged Howard to ratify the Kyoto Protocol, a measure that Kevin Rudd, the ALP opposition leader, had promised to do if he were elected prime minister in 2007. Howard lost not only his position as prime minister in the 2007 election but also his seat as an MP in the electorate of Bennelong to ALP candidate Maxine McKew, an Australian Broadcasting Corporation (ABC) journalist and a newcomer to party politics. Brett (2020) reports: "In 2013, in a lecture to London's Orwellian-named Global Warming Policy Foundation, Howard admitted that his emissions trading scheme was not evidence of a changed mind, but a pragmatic political response to a shift in public mood." As one of the more progressive Liberals in a deeply conservative party, in the end, Turnbull proved to be a thorn in the side of the climate skeptics in the Coalition, particularly Nick Minchin, Howard's finance minister (Wilkinson 2020: 71). However, Howard's last-minute effort to redeem himself by accepting the notion of an ETS did not work for him in that he lost the 2007 election to Kevin Rudd in an election in which climate change proved to be a central issue.

The ALP government and the Coalition opposition, 2007–2013

Upon becoming ALP prime minister in 2007, Kevin Rudd ratified the Kyoto Protocol, leaving the United States as the sole developed country to have never ratified it. He referred to climate change as the "great moral challenge of our generation" and convened a climate change summit with a video message from Al Gore (quoted in Grattan 2020: 470). Despite Rudd's recognition of the gravity of anthropogenic climate change, he announced on December 12, 2008, that $580 million of public money would be allocated to the expansion of coal facilities (Jowit and Wintour 2008: 1). Indeed, while serving as chief of staff to the Queensland premier, Rudd had appointed Sir Rod Eddington, a member of the Rio Tinto board, as a key business advisor on greenhouse policy. Many senior bureaucrats who had supported Howard's climate policies moved from positions in the Department of Foreign Affairs and Trade to the Rudd government's newly created Department of Climate Change in 2007 (Pearse 2009: 48). In essence, parts of the government became a branch of the mining industry. In September 2008 the Rudd government established a Global Capture and Storage Institute, with $AU100 million, to promote research into carbon capture and sequestration (CCS), with a view to creating 20 commercial plants by 2020, although in the end, this scheme led nowhere. Kevin Rudd voiced his support for the coal industry and the coal-industry/state nexus by stating that the development of "clean-coal technology" constituted a significant climate change mitigation strategy. The Victorian, New South Wales, and Queensland governments also provided big funding and policy support for CCS. However, CCS is an energy-intensive technology and unlikely to be available commercially for some time to come, if ever.

Rudd proposed the creation of an emissions trading scheme called the Carbon Pollution Reduction Scheme. Barry Jones, a former ALP minster of science, provides an astute commentary on the choice of words in the proposed legislation:

> In Australia, Labor became entranced by the use of the term "carbon pollution", and avoided references to "global warming", "climate change", or "greenhouse gases", or to explaining the science. "Carbon pollution" sounded like a clean-air campaign, a public-health measure, of great benefit to asthmatics. Who could possibly object to clean-air measures? But the term

"carbon pollution" was meaningless and unknown to science. Neither carbon or carbon dioxide are on Australia's National Pollutant Index.

(Jones 2020: 179)

As Coalition opposition leader for a short while, Malcolm Turnbull sought Coalition support for the CPRS in the Senate, all to no avail given the proposed legislation was defeated in the Senate on August 13, 2009. The Coalition argued that a decision about the creation of an ETS should be deferred until after the UNFCCC conference scheduled to meet in Copenhagen in late 2009. Barnaby Joyce initiated the Coalition campaign against an ETS and found ready support from Nick Minchin. While Tony Abbott initially had supported Howard's ETS policy and even suggested that a carbon tax might be a preferable form of carbon pricing, he quickly shifted his position. Rudd also encountered opposition to his CPRS from the Greens who viewed it as a defective climate change mitigation strategy in that it made too many concessions to industry and was weak in terms of its targets in seeking to lower emissions. The CPRS passed the House of Representatives a second time on November 17, 2009, but met defeat on December 2 again in the Senate. According to Chubb (2014: 80), the "Greens lined up with Coalition, arguing the policy was too soft on industry, particularly the coal-fired generators, and the 5 per cent emissions reduction target was too low." Turnbull lost his Coalition opposition leadership role by a single vote to Tony Abbott later in December 2009.

Rudd felt disappointed by the travesty on climate action that occurred at the 15th United Nations Framework on Climate Change Conference of the Parties held in Copenhagen on December 7–18, 2009, apparently having been spooked by Tony Abbott's campaign against any form of carbon pricing. Rudd was unwilling to go to a double dissolution on the proposed CPRS in early 2010 (Grattan 2020: 480). Rather than the initial hope to have the CPRS take effect in mid-2011, he opted to delay efforts to pass a CPRS until 2013 and when the UN COP process has achieved greater consensus on taking climate action.

Abbott, a climate skeptic of the first order, referred to the climate science as "absolute crap" in a public meeting of some 130 people in the Victorian country town of Beaufort (Chubb 2014: 78). He was a protégé of John Howard and a staunch conservative Roman Catholic. Abbott favored a 5 percent reduction of emissions by 2020 and argued that this could be achieved through "direct action," namely by planting 20 million trees, giving homeowners rebates for solar panel installations, assisting industry and farmers in restoring carbon in the soil, and encouraging Australians to make lifestyle changes to bring down emissions (Chubb 2014: 87).

The Rudd ALP government encountered massive opposition from the mining industry when Ken Henry presented a review on May 2, 2010, which proposed a Resource Super Profits Tax (RSPT) of 40 percent on profits exceeding $AU75 million per year, while proposing a lowered company tax rate of 28 percent. Henry argued that his committee "believed that the Australian public was not getting a fair return on the nationally owned minerals it was selling to mining companies" (Benns 2011: 25). Kevin Rudd refused to back down on this proposed tax and his overthrow as prime minister in 2010 was orchestrated by a $AU22-million media campaign by the large mining companies. The Minerals Council of Australia contributed $AU12.2 million to the campaign, BHP Billiton contributed $AU4.2 million, and Rio Tinto contributed $AU 537,000 (Cleary 2011: 81).

Julia Gillard, Rudd's deputy prime minister, became prime minister in a set of circumstances that may have appeared to the rest of the world as a coup d'etat. Edwards vividly captures the essence of the mining tax:

> The mining tax is widely considered the benchmark case for corporate power in Australian democracy. It is the stellar case where this battle of the giants played out in public, and climaxed with Australia's first ever deposing of a first-term Prime Minister. It was contest over how much of the benefit of the mining boom should go to the Australian taxpayer and how much should go to the mining companies. The Australian people own the minerals under the ground and the mining companies pay to extract those resources and sell them on to buyers.
>
> *(Edwards 2020: 3)*

One of the first actions Gillard took as prime minister was to whittle down the RSPT from 40 percent to 22 percent and renamed it the Mineral Resource Rent Tax (McAuley 2010). In arriving at this new arrangement, the Gillard government entered negotiations with the three largest multinational corporations in the mining industry, namely BHP Billiton, Rio Tinto, and Xstrata (later acquired by Glencore), all of which have coal mining operations. ALP Resources Minister Martin Ferguson facilitated the discussions. The revamped tax only applied to iron ore and coal companies which earned annual profits above $AU50,000 million (Benns 2011: 29). Edwards reports:

> The extent of the Labor government's capitulation became clear as the revenue collected by the next tax came in. The government attempted to conceal how much ground they had surrendered, but the numbers told the story. Over the next two years, projected revenues from the tax was progressively downgraded from $37 billion over the first four years to just $9.1 billion. Even this downgraded estimate was still too high, with the 2012/13 revenue 90% below the forecast, at just $0.08 billion, or $88 million. In the same year the mining industry income topped $200 billion.
>
> *(Edwards 2020: 30)*

Despite the propaganda campaign opposing the RSPT, public support of it continued to be high, once again illustrating the ability of a large corporation to make and break governments and politicians.

Gillard sought to defuse Abbott's attacks on her efforts to place a price on carbon by proposing the creation of a Citizens Assembly designed to deliberate on climate change policy for a year. She initiated a Multi-Party Committee on Climate Change, which produced a Clean Energy Futures plan, proposing a carbon price and support for renewable energy sources, which Coalition opposition leader Tony Abbott vehemently attacked. At a rally in Canberra in March 2011, Abbott "stood in front of placards that labelled Gillard – Australia's first female Prime Minister – a witched and the plaything of the Green leader ('Bob Brown's Bitch')" (Hudson 2019: 585).

In cooperation with Greens, the Gillard government managed to push a Carbon Pricing Mechanism (CPR) through Parliament which took effect on July 1, 2012. In the build-up to the passage of the CPR, the Gillard government circulated a booklet entitled *What the Carbon Price Means for You* to all Australian households. Jones comments upon the thinness of the booklet:

> There was not one sentence about climate change, or an explanation of why carbon pricing was being introduced. It was the weakest argument for an important national policy change since the official "Yes" case in the referendum for an Australian republic in 1999.
>
> *(Jones 2020: 183)*

An initial carbon tax of $23 per ton of CO_2 emitted was slated to increase to $24.15 in 2013–2014 and to $25.40 in 2014–2015 for the 500 top polluters. The CPR was an extremely complex scheme that included numerous compensation subsidies for both industry and low- and middle-income households. It was planned that the carbon tax scheme would transform into an emissions trading scheme in mid-2015. During its brief period, emissions reportedly had declined 7 percent according to Flannery (2020: 25). Warren reports a slightly lower drop in emissions due to a fluke of sorts:

> Emissions did fall by about five per cent during the two effect years of this short-lived carbon tax, but this was largely due to how the policy changed the generation strategy of two businesses: Snow Hydro and Hydro Tasmania. The carbon tax pushed up the wholesale price of electricity as coal generators tried to pass on the cost of the new carbon taxes. Hydro generators got paid the revenues from the price increase, but paid no carbon tax. Knowing this would occur, hydro generators withheld as much water as possible before the tax started and then increased their output during the two-year carbon tax period. This net increase in zero-emission generation reduced total emissions for the period of the tax. But it was unsustainable simply because they would have run out of spare water.
>
> *(Warren 2019: 71)*

While the Carbon Pricing Mechanism functioned during a period when electricity prices increased by $172 in the average bill, for which most households received significant compensation, the average bill had increased by $580 between 2008 and 2014 largely due to the "costs of managing customers, advertising and paying profits to the owners of the electricity companies," allowing the Coalition opposition with Tony Abbott as its leader to blame the "carbon tax" for escalating prices (Denniss 2016: 37). The ALP government did manage to strengthen its renewable energy target and create the Australian Renewable Energy Agency (ARENA), the Clean Energy Finance Corporation (CEFC), and the Climate Change Authority (CCA) (Brett 2020: 57–58).

In 2012 the Australian Government (2012: ix) released its *Energy White Paper* which proclaimed its commitment to a "clean energy economy" by seeking to "provide secure, reliable, clean and competitively priced energy to consumers while building national wealth through the safe and sustainable exploitation of our energy sources." The report projected that energy production would be boosted by coal-seam gas (CSG) and potentially shale-derived gas. Australian CSG production had increased from 2 percent to 11 percent of total gas production in the five years prior to 2010–2011 and could be expected to increase more by three CSG-to-LNG projects being constructed near Gladstone, Queensland.

Sensing a public loss of confidence in the carbon tax and staunch criticism of it by Tony Abbott, the leader of the Coalition Opposition, the ALP reinstated Kevin Rudd as prime minister in June 2013, which prompted many Australians to conclude that the ALP was in a state of disarray, a response that convinced many to vote for the Coalition in the September 2013 federal election.

At the state level, various ALP premiers supported the fossil fuel industry during the Rudd-Gillard era. When in 2007, Australian energy company Santos proposed building a liquid natural gas facility (LNG) facility, then Queensland state premier Peter Beattie praised the proposal (Reucassel 2020: 77). The federal ALP government approved the project in 2010.

The Coalition government and the ALP opposition, 2013–Present

In the background of the devastating bushfires in the Blue Mountains west of Sydney in October 2013, in response to UNFCCC head Christina Figures's observation that increasing heatwaves are likely to result in increased bushfire risk, Prime Minister Tony Abbott asserted that she was "talking through her hat" and argued that bushfires are "not a function of climate change, they are just a function of life in Australia" (quoted in Butler 2017: 5). The Coalition government which assumed power in September 2013 under Prime Minister Tony Abbott pushed for the abolition of the carbon tax in the Parliament in 2014. The federal government immediately abolished the Climate Commission, which became transformed into the Climate Council, an independent, community-funded think tank. In May 2014, it cut funding to the Commonwealth Scientific Industrial Research Organisation, the Climate Change Authority, and the Australia Energy Agency and in mid-year abolished the Carbon Pricing Mechanism. In August 2015, the federal government adopted an emissions reduction target of 26–28 percent below 2005 levels by 2030.

The Abbott government replaced the ALP's Clean Energy Package with a climate policy termed Direct Action and created an Emissions Reduction Fund designed to pay for reductions in greenhouse gas emissions through competitive bidding. Butler (2017: 36) observes that the "Clean Energy Regulator testified to the Senate that more than three-quarters of the first round of payments, for example, were made to projects that already existed – in some cases, for years." In actual fact, Abbott has engaged in considerable flip-flopping on climate change policy prior to becoming prime minister, suggesting that his real interest was using it as a way of maneuvering his way into the prime ministerial slot (Crowe 2019: 90–91). In July 2009, when Turnbull as Coalition opposition leader negotiated on the CPRS with the Rudd government, Abbott shifted from opposition to an ETS to supporting one, lest the ALP arrive at an agreement on carbon pricing with the Greens. Upon managing to be part of a spill that exiled Turnbull to the backbench in 2009 and making Abbott the opposition leader, he shifted again to opposing carbon pricing during Rudd's and Gillard's stints as prime minister. Abbott's repeated mantra became "coal is good for humanity" (Wilkinson 2020: 250). His government favored the development of a mega-mine in the Galilee Basin in central Queensland with its coal shipment going through the Great Barrier Reef and defended coal seam gas development, arguing that farmers would benefit from the construction of roads to facilitate this burgeoning industry (Cleary 2012: 108). Gina Rinehardt, a mining heiress and a multi-billionaire, obtained in May 2012 Coalition-led Queensland government approval for the first of two huge proposed coal mines, known as the GVK-Hancock mine, in the Galilee Basin (Cleary 2012: 78). In March 2015, the newly elected ALP government in Queensland paved the way for possibly additional mining in the Galilee Basin by approving the dredging of the Great Barrier Reef sited Abbott Point coal port expansion, alleging that this would be environmentally sustainable as the debris matter would be dumped at an industrial site rather on the Great Barrier Reef or Caley wetlands.

In September 14, 2015, in yet another party spill, Malcolm Turnbull replaced Abbott as the prime minister, by a vote of 54 to 44. While he attempted to create a climate and energy policy that would not split the Coalition, he made an agreement with leading Coalition members to "retain Abbott's climate and energy policies in full" (Butler 2017: 39). In December 2015 Australia became one of the 195 countries signing the UN Paris Climate Agreement. Turnbull managed to win the 2016 federal election with a one-seat majority, "making his survival dependent on the National Party – a party dominated by climate-change sceptics and deniers" (Jones 2020: 183). In the hope of having Australia meet its weak Paris Agreement emissions target, he turned to the Clean Energy Finance Corporation and the Australian Renewable Energy Agency for advice. Turnbull requested them and Alan Finkel, the Chief Scientist, and his Finkel review process to consider "pumped hydro" under his proposed Snowy Hydro 2.0 scheme, which would have entailed building "plants that would pump water uphill to store it in an upper reservoir from the plant's main dam when surplus renewable energy was available" (Wilkinson 2020: 276). However, he also proposed that coal be kept in the energy mix by building new high-efficiency low-emissions (HEHE) coal-fired plants to replace the old coal-fired plants, such as Hazelwood in the Latrobe Valley, slated for shutdown. Between 2013 and 2016, Coalition governments under both Abbott and Turnbull sought to abolish the Renewable Energy Target (RET), the Clean Energy Finance Corporation (CEFC), and the Australian Renewable Energy Agency (ARENA), but failed in obtaining Senate approval for these proposals. Grudnoff (2020: 5) argues: "Saving the RET and CEFC has resulted in reductions of at least 334 million tonnes CO_2e."

In late 2015, the Coalition with Malcolm Turnbull as its prime minister, his government re-approved the Indian-owned Adani's Carmichael coal project. Despite his repeated assertion that he takes the climate science seriously and his purported commitment to climate action, Turnbull once claimed that stopping construction of the Adani-Carmichael mine would not make "one iota" of difference in greenhouse emissions, and this action would "prevent poor people in developing countries from gaining access to affordable energy" (Denniss 2016: 17). Adani initially planned to eventually export "up to 60 million tons of thermal coal for 60 years from Queensland's Galilee Basin," resulting in massive carbon dioxide emissions (Amos and Swanm 2015: 1). This projection later was reduced to 40 million tons per annum and still later to 25 million tons per annum coal by 2020 (Quiggan 2017). Due to public pressure, Adani even more recently scaled back its plans to export an initial 10–15 million tons per annum, with a capability to expand up to 27 million tons per annum in the future (Gleeson 2018: 8). Gautam Adani is an Indian multi-billionaire who seeks to build the largest coal mine in Australia's history (Krien 2017: 14). Nearly 60 percent of India's electricity supply comes from coal-fired power plants and the government seeks to double coal production by 2020, utilizing both domestic and imported Australian coal mined by Adani (Talukdar 2016). Along with the federal Coalition government, the project was supported by the Queensland ALP government and by former ALP opposition leader Bill Shorten, who qualified his position by arguing that the mine would have to operate within certain environmental parameters.

In May 2016 the federal government requested that UNESCO remove all mentions to Australia and Great Barrier Reef in its report on the impact of climate change on World Heritage sites. In October 2016, federal, state, and territory energy ministers tasked Chief Scientist Alan Finkel with developing a blueprint for National Electricity Market (Finkel Review) that would consider the need for reliable, secure, and affordable and clean power in Australia.

When on February 8, 2016, South Australia had a heatwave, the state underwent a blackout in which 90,000 homes experienced a power blackout for 45 minutes. According to Wilkinson (2020: 277), the

> outage was caused by a series of failures involving to varying degrees the
> Australian Electricity Energy Market Operator (AEMO), an Engie- and

Mitsui-owned gas generator (which was later sued by the Australian Energy Regulator) and a disastrous software glitch from the power network's owner in South Australia.

Turnbull used the incident to attack the ALP's renewable energy policy, blaming in particular Bill Shorten. This provided an opportunity for Treasurer Scott Morrison to pass a lump of coal around Parliament to underline the government's commitment to coal. He "denounced the Labor opposition for its 'pathological, ideological fear of coal'," arguing that coal was a vital resource for economic prosperity and jobs, while the prime minister kept his head down, choosing not to look at his treasurer (Jones 2020: 183). By December 2016, a Tesla battery prevented further blackouts in South Australia.

On June 9, 2017, Alan Finkel, the federal government's chief scientist, delivered his report titled *Blueprint for the Future: Independent Review into the Future Security of the National Electricity Market*. The Finkel Review recommended a Clean Energy Target and gas generation as an interim strategy on the transition to renewable energy sources to replace coal generators that were being phased out in Australia, but the government failed to implement his recommendations. Modeling for the Review was based on a reduction of greenhouse gas emissions of 28 percent by 2030, and zero emissions by 2070. Turnbull pushed for a National Energy Guarantee (NEG) which purported to "ensure energy security while retaining a commitment to reduce carbon dioxide emissions," but it failed to obtain support from Abbott and his allies, forcing him to abandon the proposed measure (Jones 2020: 184).

The Coalition has continued this trajectory by supporting the coal mining and coal seam gas industries, a pattern that has been the practice of Australian governments in the past. In addition to supporting the coal industry per se, federal and state governments have approved the exploration of CSG sites around the country. In New South Wales, CSG exploration licenses cover most of the state, and the government there has exempted the industry from royalty payments for five years in the event mining commence. In May 2018, the Turnbull Coalition government announced a grant of $AUS443 million to the Great Barrier Reef Foundation without a tendering process. Funding fails to address climate change as a key threat to reef. In yet another party spill in August 2018, Scott Morrison, a devout Pentecostal, replaced Malcolm Turnbull as prime minister (Boyce 2019). Wilkinson reports:

> The Liberal Party leader [Turnbull] who had argued long and loud for climate science was gone. Whether he fought the climate sceptics head-on, as he did as opposition leader, or tried to appease them, as he did as prime minister, he could never defeat them. Climate and energy policy would remain hostage to the Coalition's sceptics and their allies for now.
>
> *(Wilkinson 2020: 294)*

The Morrison government has proposed using a carbon accounting loophole giving it credit for land use changes from the expiring Kyoto Protocol to fulfill its Paris Agreement commitments.

Morrison's chief of staff is a former deputy CEO of the Minerals Council of Australia, and his energy minister Angus Taylor is a long-time opponent of wind farms. In his response to the IPCC's report on the urgency of action to avoid warming of more than 1.5 degrees Celsius, Morrison declared that his government was "not held to any of [those recommendations] at all, and nor are we bound to go and tip money into that big climate fund," referring to the Green Climate Fund (quoted in Hudson 2019: 587). He expects Australia to meet its Paris targets of producing 23.5 percent of its electricity from "clean energy," while at the same time government plans to invest in new coal-fired power plants.

Although Morrison managed to retain his prime minister position in a federal election on May 18, 2019, one that many political pundits predicted would be won by Shorten, the ALP opposition leader, the 2019 federal election had an uncanny twist to it. Zali Steggall, a lawyer and former Olympian ski champion, upset Abbott in the generally safe Liberal seat of Warringah. Although politically an outsider, various determined residents orchestrated a successful campaign to oust Abbott with support from GetUp!, an online social advocacy NGO. Along with Abbott's abstention on voting on the same same-sex legislation, which easily won in a national referendum, Abbott's "perceived status as a climate change denier, his role in Turnbull's downfall, and his comments supporting paedophile [Cardinal] Pell, Warringah citizens had had enough" (Grattan and Seaborn (2019: 280).

The coal and coal seam gas industries experienced a tremendous boost due to the resounding victory of the Coalition over the ALP in the federal election on May 18, 2019. In the build-up to the election, the ALP proposed a more ambitious emissions reduction target – 45 percent below 2005 levels by 2030, compared to the Coalition's target of 26–28 percent. In contrast to other Australian political jurisdictions, this worked against the ALP in Queensland which particularly played a crucial role in providing victory to the Coalition, given that many voters there believed that an ALP government might endanger the prospect of new jobs resulting from the Adani project, despite the fact that the ALP, in contrast to the Greens, had never unequivocally opposed it. On the proposed Adani mine, Shorten, the leader of the ALP opposition, did not want to "offend the Construction, Forestry, Maritime, Mining and Energy Union (CFMEU), which provided him with the critical factional support" (Jones 2020: 185). Due to the Coalition's victory, assisted by the ALP's loss of the two seats in Queensland, Scott Morrison continued on as prime minister. Upon learning that the Coalition had won the election, Bill Shorten, the ALP opposition leader, resigned his position and was replaced by Anthony Albanese, a stalwart of the ALP left faction. Joel Fitzgibbon, an ALP MP from the mining seat of Hunter in New South Wales, "blamed Shorten for not supporting coal" (Jones 2020: 186). Although Albanese has committed the ALP to a net-zero carbon emissions target by 2050, he did not call for the keeping coal in the ground as have the Greens and many environmentalists and climate activists. Queensland ALP premier Annastacia Palaszszuk is a staunch supporter of the coal industry and particularly the Adani project, which she argues will provide needed jobs for Queenslanders.

Quentin Beresford succinctly captures the overwhelming support given to the Adani Carmichael coalmine project by the corporate-state nexus in Australia:

> The combined influence of mining barons, their peak lobby groups, the lobbyists employed by the mining corporations along with the News Corporation, the Institute of Public Affairs and the "big four" banks has resulted in the consolidation of an oligarchic model of power in Australia, especially around fossil fuels. Together these interconnected interests captured the mainstream political parties and imposed a massive expansion of the coal industry on a public increasingly worried about climate change and the fate of the Great Barrier Reef.
>
> *(Beresford 2018: 358–359)*

Although a number of coal-fired power plants have been closed over the past several years in Australia, several senior Coalition MPs would like to rejuvenate this flagging industry by nationalizing the remaining coal-fired power plants, despite opposing nationalization in general, including that of the banks (Denniss 2019: 1). While opposing subsidies for renewable energy projects, they favor subsidization of the Adani Project.

The National Party has been pressuring the Coalition government to "exclude agriculture from gashouse gas reduction targets, claiming farmers would be required to take action to reduce emissions" (Broughton and Garcia 2021: 20). However, Farmers for Climate Action (est. 2015) has opposed this move.

Although the ALP's stance on climate change mitigation, such a pledge to reduce greenhouse gas emissions at least 45 percent by 2030 against 2005 levels, while in opposition has been more ambitious than the Coalition's, by and large it has not wavered from its commitment to the coal and coal seam gas industries. In its status as the opposition in the federal parliament since 2013, the ALP has been in disarray over its climate change policy. Joshi reports:

> Members of the party like Joel Fitzgibbon and Richard Marlies have sounded out increased funding for the mining and export of fossil fuels, alongside a refusal to rule out new coal-fired power stations, as long as they're privately funded. When asked by the ABC's David Speers whether Australia will still be extracting coal and selling it to the world in the year, opposition leader Anthony Albanese simply replied "I suspect we will".
>
> *(Joshi 2020: 189)*

In seeking to explain the causes for Australia's devastating Australia's 2019–2020 mega fire season, Coalition politicians blamed it on arson, the alleged policy failures of the ALP and the Greens, poor forestry practices in limiting the extent of controlled burn, children playing with matches, cigarette embers from passing motor vehicles, exploding horse manure, a long history of devastating bushfire, and the assertion that the climate is always changing (Jones 2020: 172–173).

Despite the ALP's somewhat more ambitious emissions reduction targets compared to the Coalition, under the opposition leadership of Anthony Albanese, it appears to have backtracked on the party's support for some form of carbon pricing in the form of the Carbon Pollution Reduction Scheme during Kevin Rudd's government and later the Carbon Pricing Mechanism during Julia's Gillard's government. In his address on June 24, 2020, Albanese dismissed carbon pricing because the renewable energy costs have been declining and the potential of for carbon capture and sequestration for reducing emissions.

Linkages between the Coalition Party and the fossil fuels and mining industry are huge and include the following:

- John Kunkel, Scott Morrison's principal private secretary worked for Rio Tinto;
- Ian Macfarlane shifted from his tenure as Minister for Industry to serving as the CEO of the Queensland Resources Council;
- Angus Taylor, the Minister for Energy and Emissions Reduction, had been a consultant for the Minerals Council of Australia;
- Barnaby Joyce, a former National Party Deputy Prime Minister, has strong links with mining magnate Gina Rinehart;
- Sophie Mirabella, a former Liberal MP, now works for Gina Rinehart (Jones 2020: 188–189).

Rather than transitioning from coal to renewable energy, various Coalition politicians, such as Angus Taylor, the minister for energy and emissions, have been campaigning for a "gas-fired recovery" as a strategy for addressing the economic downturn resulting from the COVID-19 pandemic. One of the fellow travelers in this effort is former mining executive Nev Power, chair of the government-appointed National COVID-19 Coordination Committee, a group stacked with natural gas industry insiders, who have called for subsidies for new gas extraction, new gas pipelines, and new gas-fired power plants (Reucassel 2020: 77; Joshi 2020: 195). Power had been a former Fortescue Metals CEO and is presently a board member of gas company Strike Energy. Members of the group included Catherine Tanna, the CEO of Energy Australia (which owns the most polluting coal-fired power plant in Australia and has a stake in a proposed coal seam gas project), and Andrew Liveris, the director of Saudi Aramco (Greenpeace Australia Pacific 2020: 9).

Climate change denialism in Australia: The link between the corporate sector and the Coalition Party

Although many corporations, mainstream economists, and others have come to embrace some variant of green capitalism and/or ecological modernization, many other corporations, neoconservative think tanks, and conservative pundits fall into a disparate category termed *climate deniers* or *climate sceptics*. Within this amorphous grouping are several categories of bodies or people: (1) those who deny the reality of climate change; (2) those who admit its existence but claim that anthropogenic forces are not primarily responsible for it, thus attributing it primarily to natural forces; and (3) those who admit that anthropogenic activities are partly responsible but deny that it matters or argue that various social benefits may accrue from it. Morgan and McCrystal (2009: 229) argue that acceptance or rejection of mainstream climate science is closely related to an individual's political perspective.

The key features of climate change denialism are that (1) it is led by a coordinated movement (and hence is planned and organized rather than a polycentric grassroots initiative), (2) it seeks to defend a threatened ideology (e.g., the belief that capitalism is not only a productive economic system, but is characterized by a moral and beneficent agenda), (3) the funding and organizational origins of the movement are largely hidden, and (4) it seeks to sow confusion in the media and among policymakers and in the general public. Although a survey of 1,372 climate scientists conducted by the US National Academy of Sciences found that 97–98 percent of them accept the IPCC's contention that recent climate change is largely anthropogenic, reflecting an usually high level of agreement among scientists, climate deniers have clung to a set of disinformation objectives and have continued to gain "great visibility in the media, particularly after the climate summit in Copenhagen" in 2009 (Wijkman and Roeckstrom 2011: 87–88).

Another aspect of climate change denial discourse entails a strong tendency to posit the existence of colossal conspiracies and grand hoaxs perpetrated by those who express concern about global warming or climate change. In "exposing" such diabolical intrigues, climate deniers or skeptics routinely employ ad hominem attacks and direct considerable scorn at the individuals who they see as leaders of insidious campaigns of deception. In the Australian context, Tangney reports:

> Interpretations of climate change communication as more like politics (or ideology) than science (or rationality) have been demonstrated by various conservative politicians in recent years. Wielding a lump of coal in parliament in 2017, then-Federal Treasurer Scott Morrison accused left-wing opponents of "ideological, pathological" opposition to Australia's traditional energy source, at the expense of keeping the lights on. Former Queensland Deputy Premier Jeff Seeney purportedly described climate change as a "semi-religious belief" while in private sessions with Moreton Bay City Council.
>
> *(Tangney 2019: 136–137)*

Tony Abbott, a short-term Coalition prime minister, was in the habit of dismissing climate scientists or "experts" as religious zealots and thereby dismissing climate change mitigation strategies. Cardinal George Pell, the former archbishop of initially Melbourne and later Sydney and a central figure in allegations of pedophilia, asserted that environmentalists manifested a new "pagan emptiness."

The ideological center of the climate change denier movement around the world consists of a set of political conservative think tanks and public policy organizations, entities that tend to champion the virtues of the free-enterprise system or market economy, private property rights, limitations on government size and action, and a strong and globally active military. With regard to environmental issues, these organizations attempt to cast doubt on the seriousness of environmental problems of all sorts, voicing environmental concerns voiced by scientists as consisting of "questionable science, over-stated science, poorly reported science, and unwarranted statements by scientists themselves" (Michaels 2004: 5). They reject policies that are designed to improve or protect the environment, oppose enforcing corporate liability, and see environmental protection as threatening economic progress. Leading climate change denial organizations in the United States include the George Marshall Institute, the Cato Institute, the Oregon Institute of Science and Medicine, and the Heartland Institute.

In Australia, while climate change skeptics and anti-environmentalists tend to be aligned with the Liberal and National Parties, they occasionally have an alignment with the ALP. A case in point was Peter Walsh, the minister for resources (1983–1984) and minister for finance (1984–1990) in the ALP Hawke government. Kelly reports:

> A strong proponent of mining and development, Walsh believed that Labor was "infiltrated and/or unduly influenced" by green extremists in the 1980s. He was aghast when his colleague Graham Richardson experienced a green conversion and worked to convince the government to back environmental causes in return for electoral preferences. . . . Walsh was especially disgusted when Cabinet agreed to restrictions on logging in Tasmania, saying that Labor betrayed its blue-collar base, "not for any valid environmental reasons, but to appease bourgeois Left and middle-class trendoids in the gentrified suburbs of Sydney and Melbourne."
>
> *(Kelly 2019: 169)*

Walsh also rejected the findings of climate science, arguing that climate scientists were primarily seeking research grants in positing the dangers of global warming for humanity. He attended the Countdown to Kyoto conference in Canberra and argued that evidence for global warming presented at the event constituted a form of scaremongering.

After retiring from politics in 1993, Walsh went on to become the president of the Lavoisier Group, a hotbed of climate change skeptics, which is the brainchild of Ray Evans, it's secretary and the speech writer for Hugh Morgan, the CEO of the Western Mining Corporation and a pivotal figure in the Australian Mining Industry Council (later Minerals Council of Australia). Along with Walsh and Evans, the founding board of the Lavoisier Group consisted entirely of mining and energy industry notables. The Lavoisier Group convened its inaugural conference titled "Kyoto and the National Interest" in Melbourne in May 2000, "with speakers from both sides of politics, from business, for the scientific community, and two representatives from the National Farmers' Federation" (Kelly 2019: 172).

The Lavoisier Group attracted four climate change skeptics with scientific credentials, namely Bob Carter, William Kininmonth, Ian Plimer, and Gareth Paltridge. Carter was a geologist, earth scientist, and paleontologist who held adjunct academic positions at the University of Adelaide and James Cook University in Townsville. He served as a scientific advisor to the Institute of Public Affairs, a Melbourne-based ultraconservative think tank closely aligned with the Liberal Party, and a senior fellow at the Heartland Institute, a leading US climate denial think tank. While "Carter did not deny that global warming had occurred in the final decades of the twentieth century," he asserted that it was not anthropogenic and that global cooling was more likely to occur in the future (Kelly 2019: 175).

Kininmonth was a retired meteorologist who had worked for 38 years at the Australian Bureau of Meteorology and served as the head of the National Climate Centre from 1985 to 1998. He went on to establish the Australian Climate Research Institute (Kelly 2019: 176). Kininmonth asserted that climate change was due to natural forces rather than anthropogenic ones and penned a pamphlet for the Lavoisier Group titled "Climate Change: A Natural Hazard," which later was expanded into a book. Ian Plimer was initially a geology professor at the University of Melbourne and later the University of Adelaide. He served as the director of various mining corporations. In the wake of authoring *Telling Lives for God: Reason vs Creationism* in 1994, he turned his attention to what he regarded to be a secular religion, namely climate science, in *Heaven and Earth: Global Warming, the Missing Science* (2009), which sold over 100,000 copies and became "something of a bible for climate deniers both in Australia and overseas" (Kelly 2019: 177).

In May 2006 the Lavoisier Group launched Evan's pamphlet "Nine Lies about Global Warming" at Parliament House. Attendees at the event included Dennis Jensen, Russell Broadbent, and Nick Minchin from the Liberals and Martin Ferguson, Craig Emerson, and Dick Adams from the ALP (Kelly 2019: 186). Liberal Senators Cory Bernardi and Dennis Jensen launched Evan's third pamphlet titled "Thank God for Carbon" in Adelaide and Perth (Kelly 2019: 187). Bernardi went on to form his own Conservative Party in 2017. In November 2009, National Party Senator Barnaby Joyce launched the Lavoisier Group's publication *Back to the 19th Century*.

The Lavoisier Group played a pivotal role in influencing various Coalition politicians to oppose passage of Kevin Rudd's proposed CPRS legislation in 2009. With the death of Evans in June 2014, "management of the Lavoisier Group's website was transferred to the Bert Kelly Research Centre, but updates have been few and far between," but the "climate denial network lives on in Australia through other groups and individuals: the IPA, the Australian Industry Greenhouse Network, the AEF [Australian Environment Foundation] and various other lobby groups" (Kelly 2019: 193).

The Institute of Public Affairs (est. 1942) is another Australian ultraconservative think tank which has adopted a staunch climate change denial stance. It is based in

Melbourne and played a pivotal role in the creation of the Liberal Party. The IPA maintains strong links with the its more conservative members, and provides information input to various media outlets, including *The Australian*, *The Age*, the *Herald Sun*, and ABC's "The Drum," a political events TV discussion show. Its 70th anniversary dinner in Melbourne in April 2013 hosted luminaries such as Andrew Bolt, a columnist for the *Herald Sun*, Rupert Murdoch, Coalition opposition leader Tony Abbott, Victorian Liberal premier Denis Napthine, Hugh Morgan, mining magnate Gina Rinehart, and Cardinal George Pell (Pearse, McKnight, and Burton 2013: 151). In terms of climate change and fossil fuels, Pearse, McKnight, and Burton argue:

> What think tanks such as the IPA and like-minded groups effectively do is to free up the coal industry's primary lobby groups, such as the ACA [Australian Coal Association] and the companies themselves, from damaging their own credibility by being publicly identified as rejecting the scientific consensus on climate change. In this way the climate change sceptics can publicly pursue Plan A of ridiculing any action on climate change with whatever the fashionable argument of the day happens to be, while the coal companies and their lobby groups can pursue Plan B of negotiating with the government to design a response that has minimal effect on them.
>
> *(Pearse, McKnight, and Burton 2013: 152)*

The Australian, the country's only national broadsheet established by Rupert Murdoch in 1964, has served as the principal media conduit for disseminating climate denialism to ordinary Australians. Ironically, Murdoch tended to be a left-of-center media mogul until his relationship with ALP prime minister Gough Whitlam soured, prompting him to use his newspaper to vehemently campaign against the Whitlam government and help to orchestra its dismissal in November 1975 (Kelly 2019: 213). *The Australian* backed climate deniers associated with the Lavoisier Group and various other organizations have promoted the views of a string of climate deniers, including Ian Plimer, William Kininmonth, Bob Carter, and Garth Paltridge. The paper also opposed the Kyoto Protocol, emissions trading schemes, and renewable energy development (Kelly 2019: 214). Manne (2011) in an analysis of 880 *The Australian* articles between January 2004 and April 2011, found that some 700 of them opposed climate change action. Andrew Bolt, a prominent journalist based at Melbourne's Rupert-owned *Herald Sun*, also consistently promotes climate denialism in his articles.

Finally, Pauline Hanson's ultra-conservative One Nation Party staunchly fosters climate denialism, advocates that Australia withdraw from the UN Paris Agreement, and boldly boasts that it is the only Australian political party that questions the conventional climate science, including IPCC assessments (One Nation 2019).

Corporate environmentalism in Australia

As the evidence for climate change has become more apparent and acknowledged in the more progressive sectors of the mass media and on the part of even selected conservative politicians, Malcolm Turnbull being a case in point in the Australian context, in order to maintain a favorable public face, more and more corporations have adopted what might be termed *corporate environmentalism*, in some ways a more recent variant of the notion of *corporate social responsibility*.

Mikler (2013) conducted interviews with 20 senior office holders from 14 leading Australian corporations, consisting of firms involved in air travel, building materials and fixtures, heavy construction, electricity, steel, oil equipment and services, oil and gas producers, waste and disposal services, and industrial transportation. All of them accepted the findings of climate science and acknowledged that climate change constitutes one of the greatest challenge facing their companies and the need for some sort of government regulation in addressing climate change. Conversely, they expressed mixed response to the adaptation of technological innovations designed to mitigate greenhouse gas emissions.

Christopher Wright, an organizational sociologist, and Daniel Nyberg have written an illuminating book titled *Climate Change, Capitalism, and Corporations*, a truly impressive feat given that they are professors at the University of Sydney Business School and the Newcastle Business School respectively (Wright and Nyberg 2015). Although the UNFCCC, governments, politicians, and climate scientists tend to be publicly visible in the climate change discourse, private corporations tend to be less so, although increasingly so as more and more of the findings of climate science have become normalized. In light of this reality, whereas corporations were part and parcel of climate denialism, such as was the case for the Global Climate Coalition, more and more corporations have come to assert they are striving to achieve environmental sustainability and reduce greenhouse gas emissions in their business practices.

Wright and Nyberg (2015) interviewed Australian-based representatives from five corporations to unravel the basic premises of corporate environmentalism: (1) a leading energy producer that was supplementing fossil fuel generation with renewal energy sources, (2) a major financial service that was factoring a "price on carbon" into its lending policies to corporate clients, (3) a large insurance company that was measuring financial risk incurred by severe weather events, (4) an international manufacturing company that was repositioning itself as a "green" firm producing more efficient equipment and renewable energy technologies, and (5) a media company that has embraced an eco-efficiency campaign to become "carbon neutral."

Based upon their interviews with sustainability managers and consultants at these corporations, they identified three dominant identities: (1) the rational manager, (2) the green change agent, and (3) the committed activist. The rational manager emphasizes improved energy efficiency and reduced costs as environmentally sustainable practices, and believes that engaging with climate change presents new business opportunities. The green change agent is personally committed to address environmental sustainability and the climate crisis within his or her corporation, but sometimes encounters internal opposition to proposed changes. The committed activist tends to operate outside of the corporation and is involved with community environmental and climate action groups, but may experience disillusionment in trying to implement environmentally sustainable practices within the corporation.

Wright and Nyberg maintain that corporate environmentalism tends to build on the notion of ecological modernization that stresses the ability to come up with technological innovations that are environmentally sustainable. Many corporate managers view carbon pricing and eco-efficiency as "win-win" scenarios for both their firms and the environment, and they view themselves as "corporate citizens" who are promoting "sustainable businesses." Some corporations seek to pursue climate change mitigation policies by forming links with selected environmental NGOs, progressive think tanks, and even with sectors of academia.

As part of her ethnographic research on two coal mining communities in Australia, Kari Dahlgren gained entrée as a worker into a coal lobbying body established in 2011 which she designates with the pseudonym of Coal Community Conference (CCC), a subsidiary of New South Wales' Coal Lobby Organization (pseudonym), situated in Singleton, a coal mining town in the Upper Hunter Valley. She asserts that CCC "reflects broader trends toward corporate social responsibility" and takes an effort on the part of the coal industry to "appropriate the strategies of their critics, in this case, taking into account community concerns and engaging in small projects that accept the problem but do little to resolve the underlying disagreement" (Dahlgren 2019: 227–228). On the surface, CCC presented itself as neither pro-coal nor anti-coal, but as a neutral entity for expressing community concerns about the environmental impacts of the coal industry. CCC decided to embark upon an education program aimed to reach all children in primary school and later in secondary school that would take students on bus tours of nearby coal mines Dahlgren participated in a pilot program of the education program entailing transporting children from a small primary school to a nearby coal in three vehicles. She observes:

> Delaying blasting in good weather for a few hours and taking at least three managers away from other duties was a significant disruption to operations and would normally be a difficult sell to the mines, but the mine managers were surprisingly excited about this project and agreed to make the necessary sacrifices to participate.
>
> *(Dahlgren 2019: 245)*

Susanne (pseudonym), a former mining engineer and later mining manager, became an ardent fan of CCC's education project. In her opinion, mining is a fundamentally constructive and beneficial enterprise that provides many of the comforts of modernity, and in the end this turned out to be the "main message that was pushed in these school tours" (Dahlgren 2019: 246). Susanne went on to "become a national spokesperson for the mining industry appearing in a television commercial that aired repeatedly on national television" and that "was part of a broader campaign that the Australian Minerals Council – the national lobby – had recently entitled 'Australian Coal: Making the Future Possible'" (Dahlgren 2019: 247).

In New South Wales, the Coal Lobby Organization (CLO) has resisted a proposition by some environmentalists calling upon mining companies to fill in mine final voids, namely the large holes remaining in the ground upon completion of mining operations. At a CLO conference that Dahlgren (2019) attended,

> [T]he CEO of a major mining company identified the voids as the main emerging issue for the industry's community relations. Demands to fill the final voids have thus become a significant problem for the mining industry. The industry is therefore attempting to shift the narrative about them.
>
> *(Dahlgren 2019: 309)*

Nevertheless, in its attempt to smooth community relations in the Singleton area, the CCC served as a minor sponsor of rehabilitation conference at the Tom Farrell Institute based at the University of Newcastle. The CCC representatives present at the conference were frustrated by community demand for the coal industry to fill in the final voids given the costs that would be needed to complete such a program. Instead, they suggested to the community alternative options, such as using the mine voids for recreational, entertainment, tourism activities, or even developing hydropower/alternative energy (Dahlgren 2019: 314). Dahlgren observes:

> The coal lobby is aware that this framing of the voids as a potential source of entrepreneurial opportunities solves two problems in one. First, it means that they don't have to fill the voids in and they can avoid the economic costs of doing so. Secondly it speaks to demands that the mining industry assist with economic diversification and preparing the region for when it eventually exists. However, the industry itself is not actively attempting to turn the voids into business ventures of its own, or even potential venture it will then sell-on to others to operate.
>
> *(Dahlgren 2010: 315)*

Corporations involved in mining and fossil fuels industries regularly sponsor environmental conservation programs. In Central Queensland, Rio Tinto funded a program to assist the "recovery of two vulnerable turtle species on the Fitzroy catchment" (Pearse, McKnight, and Burton 2013: 110). Xstrata funded a koala habitat restoration and hair nose wombat reintroduction programs in Central Queensland.

Large corporations, including ones in the resource sector, in recent times have made concomitants to become more environmentally sustainable. Net Zero Momentum (2020), a ClimateWorks Australia project in collaboration with the Monash Sustainable Development Institute, conducted an analysis of 22 of Australia's largest greenhouse emitters in terms of their commitments to reduce their emissions. The project reports that 50 percent of companies are "fully aligned or on a pathway to net zero by 2050 for their operational emissions, which include direct emissions from the oil, gas, mining and metals processing industries and from energy purchased to power their operations" (Net Zero Momentum 2020: 5). Furthermore, Net Zero Momentum reports:

> Australia's biggest four banks have all committed to stop financing thermal coal, with ANZ, Commonwealth Bank and Westpac to cease in 2030, and the National Australia Bank to follow in 2035. Institutional investors, such as HESTA, Aware Super and BlackRock, are divesting from some fossil fuels and closely engaging with high emitting companies. Superannuation funds AustralianSuper, Cbus, HESTA, Rest and UniSuper have committed to net zero portfolios by 2050.
>
> *(Net Zero Momentum 2020: 6)*

The larger question is, however, whether these aspirational targets will ever be met and how much damage to the climate system these companies will inflict on the ecosystem even if they would be met. Perhaps such forms of corporate environmentalism constitute yet one more example of green wash.

Conclusion

President Joe Biden, a moderate Democrat, rejoined the Paris Agreement, which aims to keep the planet under 2°C, preferably 1.5°C, impose a moratorium on new oil and gas leasing on US land and water, set more areas aside for conservation, and to establish an office to serve low-income and minority communities that disproportionally suffer from air and water pollution, and establish a climate finance plan to assist developing countries in reducing greenhouse gas emissions while protecting vulnerable ecosystems and building climate resilience. While all of these measures still operate within capitalist parameters, they are much more progressive than the climate policies of former President Trump, a staunch climate denier who quickly withdrew the United States from the Paris Agreement upon assuming office in January 2017. Only time will tell whether or not Biden will prompt both Prime Minister Scott Morrison and opposition leader Anthony Albanese to take stronger climate action, given that both have refused to outline plans on how Australia can reach net-zero emissions by 2050, a policy which they claim to support.

At the same, it is important to note that the 2015 Paris Agreement is a flawed document and needs to be radically strengthened if it has any hope of achieving what it seeks to achieve. Torstad observes:

> In the Paris Agreement, self-determined targets defined through Nationally Determined Contributions (NDCs) that are not legally binding ensured an easy opt-in and low commitment costs. Second, an ambitious agreement with broad participation is likely to suffer from compliance deficits unless it contains an enforcement mechanism.
>
> *(Torstad 2020: 762)*

One study indicates that if even the current NDCs are all implemented, it would limit global warming to around 2.7°C (Rogelji et al. 2016).

Based upon 44 interviews with policymakers and policy-shapers in Australia and Norway, two countries economically reliant on fossil fuels industries, Farstad observes:

> The Australian interviews indicate that the large size of the country makes it harder to overcome interregional differences of interests and values on climate change – both as a consequence of a lack of feeling of national unity and right-wing politicians wanting to prioritise regional and constituent interests, and because of longer and poorer lines of communication. The benefits of shorter lines of communication are highlighted in the Norwegian case. With closer proximity and increased interaction between levels of government, regional fossil fuel interests have posed less of an obstacle to the creation of cross-party consensus.
>
> *(Farstad 2019: 13)*

4

CLIMATE POLITICS AT THE MARGINS

The Greens, labor unions, think tanks, environmental NGOs, and the grass-roots climate movement

Since the beginning of the 21st century, a climate movement, in part an outgrowth of an earlier environmental movement, emerged as the gravity of anthropogenic climate change has become recognized in many quarters of Australian society. Around the world, many environmental groups and Green parties have become involved in the climate movement as have other political actors, including grassroots climate action groups, socialist parties and groups, labor unions, religious bodies, and more progressive politicians. Charles Derber (2010: 205) maintains that to address the climate crisis, social movements must acknowledge the existential reality of climate change, what some term a "climate emergency," struggle for regime change within nation-states and internationally, labor, integrate personal and systemic changes, and both cooperate with the heads of state and push them to take more radical stances on systemic problems, not any easy task given vested corporate interests, particularly on the part of fossil fuels industries and motor vehicle and aircraft manufacturers and airlines. Many environmentalists, however, may not want to put labor in a position of leadership because many labour union leaders are committed to providing their members with a high material standard of living with little thought of the environmental consequences associated with it or even campaign against the end of reliance on coal and other fossil fuels. Conversely, environmental groups all too often have been willing to cooperate with corporate interests, even the more progressive ones. James Anderson observes,

> National and international NGOs, such as Friends of the Earth, Greenpeace, Oxfam, and other international aid and development organizations, play a crucial role in mobilizing scientific expertise, and some are important players in anti-globalization. But despite this progressive role, they generally occupy an ambiguous and often highly constrained position in the power game and

DOI: 10.4324/9781003202530-5

hence are vulnerable to incorporation and ideological subversion by established powers.

(Anderson 2006: 252)

Given corporations and most governments along with the United Nations have not been acting in a responsible and effective manner in terms of serious climate change mitigation, despite considerable rhetoric to the contrary, much of the collective effort must be spurred by anti-systemic movements, including a still burgeoning international climate movement that is quite variable in terms of addressing social justice or equity issues and challenging global capitalism.

The climate movement, both internationally and nationally, is a broad and disparate phenomenon that draws in part from earlier movements, particularly the environmental but also the anti-corporate globalization movement, indigenous rights movements, and even the labor movement. Countries with active climate movements include those in the Global North, such as the United States, Germany, UK, Canada, Australia, and New Zealand, and ones in the Global South, such as South Africa, Brazil, and India. Globally the climate movement can be divided into the climate movement per se and the climate justice movement. Whereas the former tends to focus on technological fixes, particularly renewable sources of energy and market mechanisms to address the climate crisis, the latter (while acknowledging a need to shift to renewable energy sources) emphasizes social justice issues and calls for "system change, not climate change." The climate movement both at the international level and within specific countries is quite fragmented, with some of its segments, such as the Climate Action Network, Greenpeace, and Al Gore's Climate Reality Project, trying to work within the parameters of global capitalism, and others, such as Friends of the Earth, Rising Tide, Climate Justice Action, and various socialist and anarchist groups, challenging it and even calling for transcending it.

In this chapter, I focus upon the Australian climate movement, within which I have functioned as a scholar-activist. Since 2008 I have been a climate justice activist, attended numerous climate rallies and climate action conferences, and worked in one capacity or another with various climate action groups, including Climate Action Moreland, the Climate Emergency Network, the Socialist Alliance, Psychology for a Safe Climate, and Al Gore's Climate Reality Project.

Attitudes of Australians toward climate change

Whereas a Lowey Institute poll in 2012 found that only 36 percent of Australian agreed that global warming is a serious and pressing issue and that steps should be taken to address it immediately, a Lowey Institute report in 2019 found that the percentage who agreed that global warming is a serious and pressing issue had increased to 61 percent and 70 percent were in favor of some form of carbon pricing (Joshi 2020: 204). In a similar fashion, the Climate Institute (2016) reported that whereas 64 percent of Australians in 2012 believed that climate change was happening, this percentage increased to 77 percent by 2016. Based on a sample of 1,756 Australians, 18 and above, 400 from New South Wales, 404 from Victoria, 400 from Queensland, 401 from South Australia, 101 from Western Australia, and 50 from Northern Territory/Australian Capital Territory, and Tasmania, Bennett (2018) reports that 73 percent of his respondents were concerned about climate change, 78 percent were concerned about climate change contributing to more droughts and flooding affecting crop production and food supply, 76 percent were concerned that climate change would contribute to more bushfires, and 70 percent agreed that the federal government needs to implement a plan to ensure the closure of coal-fired power plants and replace them with clean energy sources. Conversely, only 20 percent thought that Australia should withdraw from the Paris Agreement, and that only 15 percent of National voters did not believe that climate change was occurring.

In wake of the 2019–2020 mega-fire, the Australian Institute's 2020 *Climate of the Nation* report drew upon a sample of 1,998 Australians aged 18 years and older and consisting of 407 respondents from New South Wales, 411 from Victoria, 416 from Queensland, 407 from South Australia, 307 from Western Australia, and 50 from the Northern Territory/Australian Capital Territory, and Tasmania. It also drew from four online focus groups with 21 participants, all "female swing voters from the federal electorates of Lindsay and Macquarie in New South Wales, and Lilley and Petrie in Queensland who believe in human-caused climate change" (Quickie and Bennett 2020: 6). The study was conducted in the midst of the COVID-19 pandemic. Selected key findings are delineated:

- 80 percent of the respondents believe that Australia is already experiencing the impact of climate change;
- 82 percent of the respondents are concerned that climate change will contribute to more bushfires;
- 83 percent of the respondents support a phase-out of coal-fired power plants;
- 65 percent of the respondents think that the Australian government should stop the development of new coal mines;
- 68 percent of the respondents support a national target for net-zero emissions by 2050;
- 77 percent of the respondents agree that tackling climate change would create opportunities for clean energy development that would create new jobs and investment;
- Only 12 percent of the respondents would prefer Australia's economic recovery would be primarily powered by gas compared 59 percent who prefer that it be powered by renewable energy sources (Quickie and Bennett 2020: 1).

In terms of gender, 77 percent of female respondents as compared to 70 percent of male respondents are concerned about climate change, and 83 percent of female respondents compared to 77 percent of male respondents believe that Australia is already experiencing the impacts of climate change (Quickie and Bennett 2020:

33). In terms of generation, respondents aged 18–34 are "more concerned about climate change (85% compared to the average of 74%, and 59% for the 65 and older group." Furthermore, the 18–34-year-olds are "more likely to agree Australian governments should plan to phase out coal mining and transition to other industries (84% compared to the average of 74%, 61% for the 65 and older group)" (Quickie and Bennett 2020: 34).

Overall, various surveys indicate that the attitudes of Australians on climate change and on climate action policies are considerably more progressive than those of the two major political parties and more in line with those of the Greens and various small socialist parties or groups. It is important to note that the seriousness with which Australians take climate change will inevitably vary widely, depending upon their socio-economic status and the region where they live, something that most polls and factors do not factor in. For example, in a survey of 997 residents in the Upper Hunter Valley, a region highly dependent upon the coal industry, in 2011 anthropologist Linda H. Connor (2016: 84) and her collaborators found a relatively high degree of climate denialism or attribution of climate change to primarily natural factors. While 60 percent of her respondents agreed that the "climate is warming" and 52 percent agreed with the statement that "human greenhouse gases [are] causing warming," 25 percent agreed that the "climate is not warming" and 44 percent agreed with the statement "any warming is due to natural causes."

The Australian Greens and the climate movement

While the two mainstream parties, namely the ruling Coalition and the oppositional ALP, have consistently supported the established coal industry and the emerging CSG industry in Australia, the Greens have adopted a strong opposition to both these industries. They have embraced the "leave it in the ground" position, which they highlight on their national website. The Greens symbolize the political reality that in some ways Australia is more democratic than either the UK and the USA in that the preferential voting system in the Senate, but not in the House of Representatives, has allowed "minority and independents a better chance of being elected than under the 'first past the post' systems in some other countries" (North 2016: 12).

The opposition of environmentalists to the flooding of Lake Pedder for purposes of hydro-electric development in the 1970s served as the impetus for the of the United Tasmania Group, one of the earliest if not the first Greens party in the world. The United Tasmania Group adopted the slogan "Politics of the future versus politics of the left and right," rhetoric that alienated workers, particularly in Tasmania's logging industry (quoted in Sparrow 2005: 204). It failed initially in its bid for representation in the Tasmanian Parliament. The Lake Peddler struggle served as a prelude to the opposition to the proposal to dam the Franklin Dam. Bob Brown, a Tasmanian general practitioner, cut his teeth in the success of this campaign, which received a boost from Bob Hawke, the leader of the ALP opposition going into the 1983 federal election. In government, the ALP managed to win a case in the High Court in a five to four decision that the Commonwealth had the power to intervene in the Lake Peddler case. Brown went on to be elected to the Tasmanian parliament and later to the Australian Senate, where he assumed leadership of the Australian Greens and served as a pivotal figure in the rise of the Greens to a significant force in Australian politics federal, state, and local levels. Brown during his charismatic leadership of the Australian Greens proved to be a contradictory figure, often assuming progressive positions, but at other time adopting conservative ones. Sparrow reports:

> Greens leader Bob Brown is the most outspoken figure in federal politics, regularly addressing demonstrations and rightly praised for heckling George W. Bush in parliament. But he has, on occasion, taken some strange positions. In 2002, he toyed with supporting the privatization of Telstra [at the time a government telecommunications enterprise] in return for a ban on old-growth logging, a proposal that dismayed many Green supporters. Though the Greens opposed the invasion of Iraq, Brown supported the initial missile strike by the United States, on the grounds it might kill Saddam and thus quickly end the conflict.
>
> *(Sparrow 2005: 207)*

In the wake of the formation of various state green parties, the Australian Greens formed in 1991 and were joined by the Greens of Western Australia in 2003 (Jackson 2016: 58). The Greens began to entertain the gravity of climate change by the mid-1990s. Bob Brown and Peter Singer (1996: 14) asserted that the Greens wanted Australia to return to its commitment made at the 1988 Toronto Conference on the "greenhouse effect" which recommended a 20 percent reduction in CO_2 emissions by 2005 along with the creation of an international treaty addressing climate change. They also called upon Australia to follow the lead of various European countries that had introduced carbon taxes. Brown and Singer (1996: 14) argued that a carbon tax "should compensate low-income earners for the disproportionate burden that a carbon tax may cause them." They argued that Greens reject the dominant ethic in developed countries which promotes consumerism (Brown and Singer 1996: 44).

Bob Brown called for a plan to completely phase out coal – both for domestic use and for export – in early 2007 (Manning 2019: 253). Over the years, his political position has been more anti-materialist and anti-greed rather than explicitly anti-capitalist and over the years he has functioned as the principal voice for a green social democratic stance within the Australian Greens. Manning encapsulates the essence of Brown's charismatic leadership of the Greens in the following words:

> Brown was uniquely Green – a personification of his party. He was a fervent environmentalist with jail time to show for it, a gay man who embodied the newer identity politics, and, as an intensely proud resident of a tiny town in Australia's poorest state, had an asceticism that radiated integrity.
>
> *(Manning 2019: 467)*

Christine Milne, the second Greens Senator from Tasmania, had made climate action her forte, regularly attending UN Framework Convention on Climate Change Congress of Parties conferences. As Manning (2019: 259) observes,

> Ahead of the 2007 election she had released a landmark Greens policy, "Re-energising Australia", which included an emissions-trading scheme for power stations and heavy industry, a carbon tax for transport and a 25 per cent renewable energy target, and called for emissions cuts of 30 per cent by 2020 and 80 per cent by 2050.

In late 2009, the Greens convened a Green New Deal conference at the University of Melbourne which I attended and had an opportunity to have a long discussion with Milne.

The Greens broke into the House of Representatives, referred to as the "lower house," in 2010 when Adam Bandt won the electorate of Melbourne election in the wake of long-time ALP MP Lindsay Taylor's decision to retire (Manning 2019: 281). While the Greens have been quite successful in placing representatives in the federal Senate due to it being a chamber of proportional representation, aside from Bandt's representation in the lower house, the House of Representatives has proven to be elusive because election to it is dependent on the rule of "first past the post." Bandt argues:

> As well as growing in the Senate, a strong Greens presence in the House is a *sine qua non* of taking real political action on climate change and inequality in Australia. The Greens are not in parliament just to make up the numbers. We are not a ginger group trying to force another party to change its position. We are not a faction of any other party. Nor are we there only to secure a handful of concessions while other parties get to implement their agendas. So we need to be in the House, especially because the House leads to government, providing an opportunity to exercise power in a very different and important way compared with being in the Senate alone. Until we govern in or our own right, we must get ready to be in balance of power in *both* house, not just one.
>
> *(Bandt 2016: 187)*

The Greens maintain strong connections with the grass-roots climate movement as well as its anti-coal and anti-CSG campaigns, which serves as a counterweight to pressures exerted by parliamentary processes to become more "pragmatic." They subscribe to the notion of *economic justice*, which entails eradicating poverty and reducing inequalities in income and wealth and maintain that "achieving economic and social justice depends upon democratic participation in economic decision-making" (Australian Greens 2018).

Australian Greens differ by region of the country. Queensland and Tasmanian Greens are the most conservative and eco-centric sections of the party, putting little emphasis on environmental justice. The New South Wales and Western Australia Greens, by contrast, do stress social justice. As largely an inner-city party, the Greens have not made strong inroads into working-class outer-city suburbs and have weak connections with the union movement, although on occasion they have been recipients of modest donations from unions, including the Australian Manufacturing Union and the National Tertiary Education Union (Manning 2019: 207). Overall, as Hilier (2010) observes, "Green politics has a middle-class ideological basis" rather than a climate justice orientation. Adds Hilier (2010), "The Greens' solution therefore is not to end the system of class oppression, but to do away with materialist demands and to change our values." While many of the problems the Greens have warned about have come to pass, including the climate crisis, deforestation, and loss of human rights, the party has not been able to make significant political gains and has not made meaningful contributions to the cause of environmental justice.

Social scientists have viewed the Australian Greens in a variety of ways. Hutton and Connors (2004: 36) depict the Greens as consisting of two tendencies, the "left Greens" who subscribe to a class-based "ideological commitment to redistribution" and the "green Greens" who have a "moral commitment to social justice." Fredman (2013) compares the Greens voters with ALP voters and Coalition voters on the basis of various dimensions delineated in the 2007 Australian Election Study (AES). As Table 4.1 indicates, Greens voters are slightly less likely to be workers or employees than ALP voters but substantially more likely to be business owners than ALP voters. The lowest percentage of workers occurred among Coalition voters and the highest percentages of upper-managers and business voters occurred among Coalition voters.

In terms of union membership, 33.4 percent of ALP voters reported membership, 25.1 percent of Greens reported membership, and 15.7 percent of Coalition voters reported membership (Fredman 2013: 95–96). In terms of educational level, Fredman reports:

> [A] significantly higher proportion of Green voters than Labor voters have bachelor or higher degrees, and a significantly higher proportion of Labor voters compared to Greens voters have trade qualifications, non-trade qualifications or no post-school qualifications. In general, Green voters are more

TABLE 4.1 Occupational Composition of Voters

Occupation	ALP voters	Greens voters	Coalition voters
Workers	82.3%	76.7%	66.8%
Upper-managers	5.2%	3.8%	9.0%
Business owners	12.5%	19.5%	24.2%

Source: Fredman (2013: 94–95).

highly educated and are less likely to be engaged in skilled or unskilled blue collar labor than Labor voters.

(Fredman 2013: 99)

On environmental consciousness, Green voters scored on "average considerably higher" than ALP voters but both Green voters and ALP voters scored more highly than Coalition voters (Fredman 2013: 98). The Greens have benefited electorally from the major parties' shared perception that too much action on climate change is bad for political business. Growth and success, as well as setbacks in the 2013 federal election, have subjected the Greens to significant pressures to change (Miragliotti 2013).

After Christine Milne decided to retire from politics and thus step down as the leader of the Greens in 2015, Richard Di Natale, a Greens senator from Victoria, stepped into the leadership position. He sought to move the Greens into a more mainstream direction, a measure that made more radical Greens very uncomfortable (Manning 2019: 332). Under Di Natale's helm, the Australian Greens (2015) launched their *Renew Australia: Power the New Economy* plan, which included the following propositions:

- Achieving at least 90 percent renewable energy generation and the doubling of energy efficiency by 2030;
- Establishing a new $500 million government authority called Renew Australia which would drive the planning and transition to a new clean energy system by leveraging $5 billion of construction in new energy generation over next four years;
- Creating a $250 million Clean Energy Transition Fund to assist coal workers and communities with the transition to clean energy;
- Implementing pollution intensity standards to enable gradual, staged closure of coal-fired power stations, starting out with the Hazelwood coal-fired power plant;
- Creating an energy market which would continue to be a mix of private, public, and community infrastructure.

By and large, the Greens can be characterized as a green social democratic party. While it has some eco-socialists and eco-anarchists who seek to transcend capitalism both worldwide and even in Australia, they tend to be sidelined within the party. An interesting case study was the emergence of Left Renewal within the Green Party of New South Wales in late 2016. Left Renewal drew upon the four pillars embodied in the Australian Greens founding premises, namely ecological sustainability, grassroots democracy, social justice, and peace and non-violence. Here is a condensation of Left Renewal's positions on these four pillars:

- "That our struggle for social justice beings us into irreconcilable conflicts with the capitalist mode of production, and all other forms of class society. This requires us to take a strong stance on the struggle of the working class";

- "We will embody in our organization the value of participatory self-management whereby everyone has a say in the decisions which affect them and the resources on which they are dependent in proportion to the degree to which they are affected";
- "That capitalism is a violent and antagonistic relation between worker, and those who exploit them";
- "We need an internationalist perspective on climate justice that recognises that we are all part of the global ecosystem whatever our ethnicity, gender, or sexuality, and that we can only resolve climate change by achieving climate justice."

Manning reports:

> Under sustainability, the manifesto rejected green-capital partnerships and market solutions such as consumer taxes, arguing that "the interests of the environment, and the working class, should be never be pitted against one another. This means working toward a 'just transition' in collaboration with and alongside the working class in order to uncover a renewable world".
>
> *(Manning 2019: 350)*

Lee Rhiannon, a long-time social activist, a former member of the Socialist Party of Australia, a former Green parliamentarian in New South Wales, and a former Greens senator from New South Wales, was loosely aligned with Left Renewal; however, she was shunted aside from her leadership role in internal party machinations (Crowe 2018: 86). To her credit, Rhiannon was a staunch advocate of a "just transition" for workers that would be displaced from jobs in the fossil fuels industry. She argued that "developing ecologically sustainable industries in manufacturing, energy and transport infrastructure, and promoting energy-efficient construction and retrofitting will create massive numbers of jobs" (Rhiannon 2012: 28). However, Richard Di Natale followed by Bob Brown and Christine Milne came out in opposition to the radical stances adopted by Left Renewal, and shortly thereafter the group went into abeyance. Milne came out of her retirement and suggested that New Left Renewal leave the Green Party and create its own party and that federal Green Senator Lee Rhiannon and New South Wales Greens Party leader David Shoebridge assist them in making an exit (Greenland 2017).

Adam Bandt served a stint as deputy leader of the Greens under Christine Milne and became the Greens leader after Richard Di Natale stepped down from his term as leader in 2019. He was a labor relations lawyer who had developed strong ties with the union movement. Manning reports:

> Bandt overhauled the party's Renew Australia policy and shifted the focus to a steady phase-out of coal exports over the 12 years to 2030. . . . Yet Australia is the world's largest exporter of coal, and, as Bandt explains when

he outlines the policy, if the emissions from burning that coal overseas are included, the country becomes the sixth-largest greenhouse-gas emitter in the world.

(Manning 2019: 393)

The main points of the Greens *Renew Australia 2030* proposed the following policies:

- Phasing out coal, including coal exports, and replacing it with a clean energy export industry;
- Requiring coal companies to pay for the rehabilitation of the land so that it can be utilized for housing, agriculture, and parkland;
- Encouraging billions of investment and create tens of thousands of jobs;
- Setting a 100 percent renewable energy target by 2030;
- Creating a publicly-owned, non-profit, energy provider;
- Developing clean technologies in Australia and exporting them to the rest of the world;
- Restoring a price on carbon;
- Investing $100 million in an Indigenous and remote communities power fund, administered by ARENA;
- Rejecting natural gas as a transitional energy source;
- Promoting an "Electric Vehicle Revolution" (Australian Greens 2020a).

By and large, while a vast improvement over the climate policies of both the Coalition and the ALP and taking the bold step of moving toward public ownership of energy production, *Renew Australia 2030* adopts a green social democratic agenda and draws upon an ecological modernization paradigm, both of which are consistent with capitalist parameters.

Bandt is clearly the most progressive federal leader the Australian Greens have ever had, more so than Bob Brown, Christine Milne, and particularly Richard Di Natale. Furthermore, he is probably the most progressive as well as an intellectual politician in the federal Parliament, although Lee Rhiannon, as a former Greens senator from New South Wales, had probably rivaled him for this distinction during her time in Parliament. Bandt has grappled with Marxism and socialism over the course of his life. As an undergraduate student at Murdoch University in Perth, he reportedly rose to the leadership of the Left Alliance, a student group aligned with the Communist Party of Australia. Bandt read both Hegel and Marx in German and did a PhD thesis on the foundations of law under the supervision of Andrew Milner at Monash University, who insisted that his student was a Marxist, at least in the theoretical sense. Bandt has looked to the partial success of anti-austerity and anti-capitalist parties overseas, particularly Syriza in Greece and Podemos in Spain (Manning 2019: 398). He maintains that regulating capitalism is the only viable way to tackle climate change (Manning 2019: 399).

Since Bandt's accession to the Australian Greens' leadership, he has joined United States Democratic Congresspersons Alexandria Ocasio-Cortez and Ed Markey and others around the world in calling for a Green New Deal. In 2020 and the midst of the COVID-19 pandemic, Bandt asked input from Greens members on the development of an Australian Green New Deal based upon the following premises:

- A government-led plan of massive investment and action to develop a "clean economy and a caring society";
- Overcoming "supercharged economic-inequality, chronic unemployment and the environment and climate emergencies";
- A just transition from fossil fuels to clean energy;
- Transforming Australia into a "renewables powerhouse";
- Empowering Indigenous people and ensuring that "marginalized communities have a say in directing their own future" (Australian Greens 2020b).

Overview of the Australian climate movement

In frustration at the lack of action on the part of the Coalition (Liberal-National parties) under Prime Minister John Howard from 1996 to 2007, at the beginning of the present century, a growing number of Australians began to join local grass-roots climate action groups (CAGs). These groups, along with various environmental NGOs, regional action groups, special climate action groups, and socialist groups, have constituted the most significant new social movement in Australian society.

Mark Diesendorf (2009: 7), an energy analyst and a climate activist, delineates several types of *climate movement organizations* (CMOs) in Australia: (1) grass-roots groups, (2) "larger environmental and social justice NGOs for which climate action is one program," and (3) "some student, faith, trade union, business and professional groups." He observes that CMOs may operate within coalitions and alliances with "influential organisations, such as large businesses, trade unions, and professional organisation" (Diesendorf 2009: 140). While Diesendorf's scheme is quite comprehensive, it overlooks various key players in the Australian climate movement and needs updating to include some new key players.

I propose an alternative and more updated scheme depicted in Figure 4.1, which distinguishes between actors in the climate movement that tend to operate in two broad layers. This scheme illustrates the evolutionary process that social movements often go through if they achieve some level of institutional success. While some might argue as to which groups in the climate movement operate within mainstream civil society and which operate at the grass-roots level, some actors in the climate movement tend to be more bureaucratic and procedural while others tend to be more spontaneous, egalitarian, and participatory. Nevertheless, even groups at the grass-roots level may exhibit tendencies toward bureaucratization and compromise with the powers-that-be.

Mainstream civil society
- Think tanks
  Climate Council
  Australia Institute
  Breakthrough
- Unions
  Australian Council of Trade Unions
  Australian Manufacturing Workers Union
  Construction, Forestry, Mining, Electrical Union
  National Union of Workers
  National Tertiary Education Union
- Peak environmental NGOs
  Australian Conservation Foundation
  State Nature Conservation Councils
  Greenpeace
  Friends of the Earth
- Grass-roots climate movement
  Local climate action groups (CAGS)
  Special climate action groups
  Australian Youth Climate Coalition
  Lock the Gate
  Stop Adani Alliance
  Students Strike 4 Climate Action
  Extinction Rebellion
  Socialist parties and groups

FIGURE 4.1   Layers and Actors in the Australian Climate Movement

The listing in Figure 4.1 is not necessarily complete and subject to change as some groups become dormant and even defunct and others appear. For example, whereas groups such as Safe Climate Australia, Climate Emergency Network based in Victoria, Rising Tide based in Newcastle, New South Wales, were quite active around 2010, they have largely become dormant or defunct (Burgmann and Baer 2012). Conversely, Students Strike 4 Climate Action and Extinction Rebellion are relative newcomers to the Australian climate movement scene.

Mainstream civil society

Numerous groups and organizations operating within the corridors of mainstream civil society as well as occasionally in parliaments and local councils, and within the view of mainstream media, have come to address climate change. In this section,

I examine some of these, namely labor unions, think tanks, and environmental non-government organizations (NGOs).

Labor unions

The late Jack Mundey, the famous leader of the Green Bans campaign in Sydney during the early 1970s and a Greens member, stated in 2016:

> The union movement now could be a real, important instrument for the whole ecological struggle, because the whole question of global warming, the whole question of the environment, is one the agenda forever more, it's not going to go away, and I believe that the union movement should become more and more involved.
>
> *(Manning 2020: 406)*

Mundey, along with Bob Pringle and Joe Owens, had led the New South Wales branch of the Builders Labours' Federation (NSWBLF) in imposing 42 green bans on the planned destruction of homes and heritage buildings to make room for office buildings (Burgmann and Burgmann 1998).

Labor unions, particularly through their umbrella organization the Australian Council of Trade Unions (ACTU), have historically had a close connection with the ALP, but this weakened as the ALP began to adopt neoliberalism in the early 1980s. As early as 1991 the ACTU adopted a "Policy on Greenhouse" that demanded the immediate implementation of all available low-cost measures to reduce greenhouse gas emissions; supported international negotiations on a climate change convention; called on governments to fund research into energy efficiency and renewable energy; acknowledged that reduction of emissions would require a mixture of pricing, taxation, and regulatory measures; argued that a price mechanism alone would adversely affect low-income and that regulatory and planning measures would be more efficient and effective; and called on federal and state governments to develop a national approach to encourage improved performance in energy production, distribution and use – in power stations, vehicles, buildings, plant and equipment and domestic appliances (ACTU 1992).

In 1997, a union/environmental group called Earthworker was established with sponsorship from Victorian Trades Hall in Melbourne (Burgmann 2013: 135). It developed a Solar, Wind and Water Alternative Energy Plan and urged the labor movement to push for green jobs and Australians towards a sustainable future.

In February 2007 the ACTU convened a national Union Forum on Global Warming and Sustainable Energy to develop strategies for mitigating climate change (Burgmann 2013: 137). In 2008 the ACTU in collaboration with the Australian Conservation Foundation released a report titled *Green Gold Rush*, which asserted that green industries could create about 500,000 jobs by 2030, thus paving the road to Australia's future economic prosperity. In August 2009 the ACTU launched a Union Climate Connectors program to provide a forum for its members to engage in climate change action. The ACTU supported the Rudd government's CPRS. The Australian Manufacturing Workers' Union (AMWU) released a report in 2008 titled *Making Our Future: Just Transitions for Climate Change Mitigation*. It views the climate crisis as an opportunity for industry rejuvenation and growth. The AMWU advocates a transition from existing manufacturing industries to green industries in keeping with the development of a green capitalist economy.

If any union had reasons to be supportive of climate change denialism, it would be the Construction, Forestry, Mining, and Energy Union (CFMEU), concerned as it was with protecting workers in the sectors most responsible for greenhouse gas emissions. This has not for the most part been the case, but there have been some exceptions. The CFMEU has long acknowledged the seriousness of climate change and the role of fossil fuels, but it aims to ensure that the necessary process of reducing emissions does not disregard the economic security of workers and their rights to alternative employment options. Its "Climate Change Policy" in September 2007 emphasized that burning Australian coal contributed to climate change (CFMEU 2007). While Tony Maher, the CFMEU national president, criticized the Greens for not having supported the Rudd government's CPRS, he expressed criticism of Rudd as well, arguing that Rudd had kept the Greens at a distance during negotiations for the proposed legislation (Chubb 2014: 82).

During the period in which climate activists campaigned for the closure of the Hazelwood coal-fired power plant in the Latrobe Valley, International Power sought support from the company's employees and CFMEU workers. Since the Coalition returned to power in 2013, CFMEU, with some backing from the ALP, has been engaged in a "fight for our living standards" and rolling back the insecure workplace of the past four decades. A CFMEU state representative in Moranbah, a coking coal town in the Upper Hunter Valley of New South Wales, in his interview with anthropologist Kari Dahlgren suggested to her that casualization of the workforce and labor dislocation through long-distance commutes to the surrounding mine sites threaten the economic viability of the community more than the end of coal production per se:

> There is acceptance globally that climate change is real. It took me a while to get there, I admit. But at the higher levels you get to understand it. . . . None of our mines have shut because of climate change: they've been shut because of the oversupply of coal causing the price to collapse, and then even if they don't shut, the company uses the lower price as an excuse to go and cut "costs." That means our entitlements. We see economics as a bigger pressure than climate change.
>
> *(quoted in Dahlgren 2019: 91)*

While the "jobs and environment" initiatives proposed by the union movement do not make any significant challenges to global capitalism, they can if revived assist to break down the historical divisions between unionists and environmentalists which have strengthened those who profit from the exploitation of both labor and nature. While CFMEU workers at Hazelwood were split on whether or not to support the proposed CPRS and the shutdown of the plant, "in March 2011 the local CFMEU took the extraordinary step of passing a resolution recognising that carbon pricing was inevitable" (Chubb 2014: 85). Another union group, the Gippsland Trades and Labour Council (GTLC), along with the Latrobe Council and Regional Development Australia Gippsland called for a "just transition" in which the government would provide funding for displaced workers at coal-fired plants in the Valley to transition into other forms of employment (Chubb 2014: 85–86). Darren Chester, the National Party MP for Gippsland, asserted "that the unions were threatening the wellbeing of their members by not opposing government plans" (Chubb 2014: 86).

The CFMEU has argued that Australia's coal industry could learn from Germany, where, in less than 30 years, some 130,000 coal miners left the coal industry in a smooth manner through voluntary redundancy packages into new jobs, with only a few hundred workers ending up unemployed (Manning 2019: 405). Much to the chagrin of climate activists, the CFMEU in Queensland asked federal election candidates in 2018 to sign pledges supporting the Adani coal project in the Galilee Basin and opposed the Stop Adani convoy organized by Bob Brown. Not all CFMEU leaders in Queensland agreed with the union's stance on Adani, with one senior official "warning the state government not to fall for inflated job promises and calling Adani a 'corporate carpetbagger'" (Manning 2019: 407). The National Union of Workers called on the labor movement to "think big and take the lead" on climate action (quoted in Manning 2019: 407).

By and large, the union movement in its promotion of green jobs has opted to operate within a green capitalist framework. Goods observes:

> [T]heir response to the broader environmental problems and the potential solution of green jobs has so far been unquestioning of the continued operation of the treadmill of production and the purpose of human labour within the treadmill. . . . The diverse positions within and between unions undoubtedly [reflect] the difficult and paradoxical choices that union leaders and members face. Nevertheless, the solution of fusing labour with ecological sustainability needs to be much more than process of adaptation. It is unlikely to succeed unless it confronts the ecologically unsustainable side of capitalism and is integrated into the broader processes of class struggle and the redress of social inequality. The labour movement should therefore take the opportunity to push the current promotion of green jobs in a progressive direction by challenging "how nature is produced", why and for whose benefit.
>
> *(Goods 2011: 64–65)*

In contrast to most other trade unions that tend to support the ALP, the leadership of the National Tertiary Education Union (NTEU) has tended to favor the policies of the Greens. For example, in terms of the NTEU scorecard on the policies on climate change and energy policy of the Liberals, ALP, and Greens, Michael

Evans (2019: 27), the NTEU National Organiser, gave his highest rating to the Greens, stating: "The Greens have the most progressive policies of the major parties around climate change and sustainable energy production." In 2020 the NTEU National Council directed the NTEU to develop a comprehensive climate justice campaign (Barnes 2020).

Think tanks

The main think tank touching upon climate change-related topics in Australia is the Climate Council, formerly the Climate Commission created by the Rudd ALP government but dismantled by the Abbott Coalition government. This section also touches upon the work of the Australia Institute, a progressive think tank based in Canberra, which examines a wide array of socio-economic issues, including energy policy and climate change policy. Breakthrough based in Melbourne focuses on various dimensions of what it terms a *climate emergency*.

Climate Council

The Climate Council started out as an ALP governmental body called the Climate Commission, with Tim Flannery as its director, which was immediately abolished by Tony Abbott when he became Prime Minister. However, the Climate Commission quickly transformed itself into the Climate Council and has funded itself with donations from concerned Australians. Except for one person, all of the Climate Commission's commissioners joined the Climate Council as councilors or board members. Flannery reports that since the demise of the Climate Commission,

> [T]he Climate Council has gone from strength to strength, and today it employs many more staff than the Climate Commission did, as it continues to provide up-to-date scientific information on climate. It has established the Climate Media Centre, which gives voice to farmers, doctors and others concerned about climate change; the Cities Power Partnership, which works with local governments to get action on climate; and the Emergency Leaders for Climate Action, all of which have been highly effective. The Climate Council has played a crucial role in creating climate awareness: today 84% of Australians want climate action – testimony to the fact that, provided with sound evidence, Australians are willing to take action.
>
> *(Flannery 2020: 31)*

The Climate Council has published numerous reports on a wide range of climate change-related issues, including climate science, the impacts of climate change on Australia, global efforts to reduce greenhouse gas emissions, and the potential of renewable energy. It has also convened numerous public meetings around Australia. In a report highlighting the environmental problems associated with proposed coal extraction in the Galilee Basin, the Council (2015) called for no new coal mines, phasing out existing coal mines and coal-fired power plants, and scaling up renewable energy to power Australia. In its report on green jobs, the Climate Council (2016b: 15) argued if Australia would draw 50 percent of its electricity from renewable energy, this would result in nearly 50 percent more jobs in the renewable energy generation sector than the fossil fuels electricity generation sectors. Furthermore, the same target would result in an additional 28,000 jobs nationwide, with 70 percent of them being from installing rooftop solar panels, 17 percent working in wind farms, 9 percent in solar plants, and 3 percent in combined cycle gas turbines (CCGT) (Climate Council 2016b: 20). The Climate Council (2019a) also published a report which discusses the federal government's attacks on climate science, its failure to meet greenhouse gas emissions reduction targets, its failure to effectively develop renewable energy, its undermining of electric vehicles, and the closure or reduction of funding for climate-related programs.

The Australia Institute

The Australia Institute is a progressive think tank established in 1994, which publishes numerous reports on a wide array of socio-political, economic, and environmental issues (see Aulby and Ogge 2016). Clive Hamilton served as the founding director of the Australia Institute with his collaborator in the authoring of the book *Affluenza*. Richard Denniss took over the directorship when Hamilton assumed an academic position at Charles Sturt University in early 2008 (Hamilton and Denniss 2005). Through its publications, the Australia Institute seeks to contribute to a "more just, sustainable and peaceful society" and "both diagnose the problems we face and propose new solutions to tackle them" (Australia Institute n.d.). Over the years the Australia Institute has published numerous briefs and reports on climate change-related issues, including the Carbon Pollution Reduction Scheme and the Carbon Pricing Mechanism, the coal industry, governmental climate policies, and so on. The Australia Institute proposed a Genuine Progress Indicator as an alternative to Gross Domestic Product that "took into account resource depletion and the costs of economic growth" (Robinson 2019: 133).

Breakthrough

The National Centre for Climate Restoration seeks to develop "critical thought leadership to influence the climate debate and policy making" (Breakthrough n.d.). Its research areas include climate impact and risk assessment, rapid emissions elimination, large-scale greenhouse gas drawdown, and innovations to prevent harm from climate change. In some ways as a promoter of the climate emergency framework in Australia, it is a successor of the Climate Emergency Network, a now-defunct climate action group. Its primary movers have been David Spratt, who serves as research director; Ian Dunlop, a former oil and gas industry executive, a former chairperson of the Australian Coal Association, and a Club of Rome member; and Philip Sutton, the manager of Research Strategy for Transition Initiative.

Breakthrough periodically publishes reports about the gravity of the climate emergency and security risks emanating from it. Since 2014, it has also convened periodic conferences and events and played a key role in other conferences, many of which I have attended. In Chapter 5, I provide a first-hand account of the 2020 National Climate Emergency Summit, of which Breakthrough served as the key organizer.

Environmental NGOs

Environmental NGOs have experienced an uphill battle to influence climate policy in Australia. McDonald (2016) asserts that they for the most part have been unsuccessful in shaping government climate policy and have also failed to turn the tide of public opposition to climate change mitigation strategies in Australia.

The Australian Conservation Foundation

The Australian Conservation Foundation (ACF) was established in 1966 "when a group of scientific, public service, business and political decision makers joined forces in the hope of establishing a body which would campaign for the protection of Australia's environment" (Lancaster 2012: 248). ACF lobbies governments, government departments, and industry; educates the public through its many publications; maintains a high media profile; and endeavors to expand its membership and support base. It has been involved in numerous campaigns over the past several decades, including:

- Protection of large portions of the Mallee in Victoria;
- Creation of the Kakadu National Park in the Northern Territory and marine parks in Victoria;
- Protection of the Franklin River in Tasmania;
- Banning of mining in Antarctica;
- Protection of the Murray-Darling River Basin from further degradation; and
- Development of the Greenhome and Climate Projects to urge Australians to make lifestyle changes mitigating greenhouse gas emissions (Lancaster 2012: 249).

For several years beginning in 2008, ACF had been a member of the Southern Cross Climate Coalition, now largely defunct, which also included the Australian Council of Trade Unions, the Climate Institute, and the Australian Council of Social Service. Essentially, it looks to government regulation of business to enforce ecological standards. ACF periodically publishes reports on various climate change-related issues, such as Australia's top "climate polluters" (Australian Conservation Foundation 2015.) and the federal government's environmental approval of the Adani Carmichael coal mine project (Australian Conservation Foundation, n.d.). While ACF has identified and condemned the ten corporations that contribute nearly one-third of Australia's greenhouse gas emissions, as Paul (2016: 73) observes, it "says nothing about cars, trucks, and planes as the biggest source of pollution in Australia." For several years, under the leadership of its director Don Henry, it maintained a close connection with Al Gore's Climate Reality Project which later, through Henry's efforts, affiliated with the Melbourne Sustainable Society Institute at the University of Melbourne.

Greenpeace Australia

Greenpeace Australia became part of Greenpeace International in 1987, with its first campaign opposing uranium mining in Australia. It has also "campaigned against the destruction of Indonesia's rainforests, overfishing, whaling, the cultivation of GM [genetically modified] crops and nuclear power and waste" (Lancaster 2012: 2012).

While not being explicitly anti-capitalist, Greenpeace Australia tends to be more progressive and more likely to engage in direct action than ACF. Greenpeace Australia has maintained a steady stream of anti-coal protests, sometimes on its own and at other times with other groups. One of Greenpeace's international campaigns has been to stopping climate change by phasing out fossil fuels and replacing them with renewable energy, particularly solar and wind. It led the way during the Howard government days of anti-coal campaigning when on July 27, 2005, teams of its activists occupied a 2.5 million ton coal stockpile and one of four huge coal loaders. On September 2, at the start of the 2007 Asia-Pacific Pacific Economic Cooperation (APEC) meeting hosted by Coalition Prime Minister John Howard in Sydney, 12 Greenpeace activists painted the message "Australia Pushing Export Coal" on the side of the coal ship *Endeavour*, and unfurled a large banner in Chinese calling on China to be cautious of Howard and George W. Bush's attempt to sabotage the Kyoto Protocol. In August 2011, five Greenpeace activists were arrested and pleaded guilty to charges of blocking a BHP Billiton coal train from the Mt. Arthur coal mine in the Hunter Valley of New South Wales to the coal port at Newcastle. They were demanding BHP Billiton pay its fair share toward helping Australia develop "clean energy" for its future. In terms of the coal industry, Greenpeace has adopted a strategy of disrupting and delaying key projects and infrastructure "while gradually eroding public and political support for the industry and continually building the power of the movement to win more" (Hepburn, Burton, and Hardy 2010: 3).

Greenpeace maintains Australia is well-positioned to implement renewable energy resources, such as solar photovoltaic, concentrating solar thermal, wind, wave, and geothermal energy. It also strongly opposes carbon capture and sequestration. Greenpeace opposes the expansion of Australian coal exports, particularly the ones to be situated on the Great Barrier Reef coast of Queensland (Greenpeace Australia Pacific 2012a, 2012b). It has joined forces with the Gomeroi Indigenous community and various other environmental groups in opposing the development of the Whitehaven Coal Company's Maules Creek mine situated, reportedly the "largest new open cut coal mine currently under construction in Australia" (Greenpeace Australia Pacific 2014: 3).

Greenpeace's collaboration with the Gomeroi Indigenous community expresses an ongoing commitment by Greenpeace to environmental and climate justice, a stance seen especially in work by Greenpeace's support of Pacific island communities fighting the elite polluters responsible for a significant portion of climate change. The Gomeroi Tribal Nation Secretariat (2012) maintains that the Gomeroi people hold inherent rights in their Indigenous area and "recognize our guardianship and ownership rights in country, including our unique responsibility to care for land and water, the ecosystem and places of cultural significance." While Greenpeace is best known for its civil disobedience in environmental action, climate justice remains an ongoing, if less vocalized, concern.

Friends of the Earth Australia

Friends of the Earth Australia (FoE) formed in 1974 on French Island, the site of a proposed nuclear reactor in Westernport Bay, Victoria. Of all the players in the mainstream civil society sector of the Australian climate movement, Friends of the Earth Australia is the most radical. Burgmann (2003: 192–194) characterizes Friends as a "radical, egalitarian, unbureaucratic organisation, committed to sustainable activism on the ground" and the "most radical of the principal [Australian] green organisations, and participated enthusiastically" in demonstrations against the World Economic Forum conference in Melbourne.

Friends of the Earth International has experienced tension between its branches in the Global North and those in the Global South. Doherty and Doyle report:

> Unlike other major environmental NGOs, such as WWF and Greenpeace, FoEI seeks system-wide change, and has made political economy a central plank of its strategy. For many FoE groups in the global South, capitalism is the main source of multiple forms of domination that directly, often violently, conflict with surviving pre-capitalist modes of existence. Capitalism is therefore to be resisted and replaced as a matter of survival. . . . In the global North, there was a reluctance to use anti-system language, and in the early stages of FoEI's internal debates North groups emphasised holding particular corporations to account. For many in the South, corporate accountability sounded too soft and side-stepped the need of broader critique of TNCs and capitalism.
>
> *(Doherty and Doyle 2018: 1064)*

While Friends of the Earth International has adopted a stronger Southern orientation, particularly on the part of its branches in the Global North, each national group retains a high level of autonomy. Friends of the Earth Australia has maintained a vigorous Climate Justice Campaign, particularly directed toward South Pacific Islanders on low-lying islands. In early 2009, it initiated a Climate

Displacement Coalition that sought to lobby the Australian government to assist climate refugees to relocate.

Friends of the Earth Melbourne has been for several years involved in a campaign for a just transition in the La Trobe Valley. Perhaps in the spirit of political diplomacy, it acknowledges various progressive policies and infrastructural projects embarked upon by the Victorian ALP Andrews government, including:

- Implementation of the Victorian Renewal Energy Target (VRE) and the setting of targets for 2020, 2025, and 2030;
- The creation of the first-ever Australian permanent ban on fracking, and the promise to embed it in the Victorian constitution;
- The preservation of a moratorium on onshore conventional gas drilling;
- The strengthening of the Victorian Climate Change Act and a commitment to a target of zero emissions by 2050 in legislation;
- The commitment of $1.3 billion to a solar homes program, which aims to provide half price solar panels to 650,000 households;
- A commitment to operate the Melbourne suburban train and tram networks on renewable energy;
- The creation of a $266 million La Trobe Valley transition package upon Hazelwood's closure (Friends of the Earth Melbourne 2019: 5).

Friends of the Earth Melbourne proposed a transition from fossil fuels to renewable energy that called for the following measures:

- The creation of a Just Transition Authority to implement a just transition;
- The creation of secure union jobs as a key element of the transition;
- The adoption of regional planning as opposed to a "plant by plant" approach;
- The creation of the world's largest offshore wind farm;
- The empowerment of communities through inclusion in the state planning process;
- The adoption of a Green New Deal approach (Friends of the Earth Melbourne 2019: 7–10).

Friends of the Earth Melbourne also recommends numerous specific initiatives to be implemented in Victoria, including funding for new public housing, energy efficiency in homes, a solar homes program, powering public transportation with renewable energy; the creation of a high-speed rail line between Melbourne and Sydney; the utilization of eco-cement in building construction; public ownership of renewable energy generation; powering the Portland aluminum smelter with 100 percent renewables; support for the Earthworker Cooperative project in the manufacture of renewable energy appliances and components, and the development of native forest carbon dioxide sequestration (Friends of the Earth Melbourne 2019: 11–23).

The grass-roots climate movement

The grass-roots climate movement, which for the most part operates below the radar of mainstream media, includes numerous climate action groups (CAGs), socialist parties and groups, the Australian Youth Climate Coalition, the Stop Adani! campaign, and most recently the student strike climate movement, and Extinction Rebellion. Whereas around 2009 and 2010, there existed a vibrant national climate action network and a few state-based climate action networks, they are now essentially defunct (Burgmann and Baer 2012: 262–273). Whereas around this time, there were some 100 or perhaps more CAGS, many of them have become either dormant or defunct but others, such as Climate Action Moreland and Climate Action Darebin in Victoria, continue to thrive.

In frustration at the lack of action on the part of the Coalition (Liberal-National parties) government under Prime Minister John Howard from 1996 to 2007, around the beginning of the present century, a growing number of Australians began to join local grass-roots climate action groups (CAGs). These groups along with various environmental NGOs, regional climate action groups, special climate action groups, and socialist groups, for a time, constituted the most significant new social movement in Australian society, although one that went for several years into a lull for complex reasons, including the passage of a Carbon Pricing Mechanism in 2012 that the Coalition government dismantled in 2014. Nevertheless, the nation-wide anti-coal and anti-CSG campaigns managed to preserve much of the earlier dynamism of the larger climate movement. These campaigns have included groups such as the Green Party, Rising Tide, Greenpeace, Friends of the Earth, Climate Emergency Action Network in South Australia, the Socialist Alliance, numerous local climate action groups (CAGs), Lock the Gate in Queensland and New South Wales, and Stop Adani, a nation-wide coalition (Burgmann and Baer 2012). These campaigns have called for the closure of coal-fired power plants and the cessation of coal mining, CSG exploration, and coal exports and expressed vehement opposition to carbon capture and sequestration. While Greenpeace activists John Hepburn, Bob Burton, and Sam Hardy (2010: 4) admit that the Australian anti-coal movement is "fragmented and under-resourced," they also assert it is "growing rapidly due to increasing impact of coal mining and coal seam gas industry on water resources, agricultural land, and rural communities."

To a large degree, the climate movement constitutes a subgenre of the environmental movement, although it overlaps with various other social movements. The climate movement in both its international and national manifestations, including the one in Australia, is broad and disparate and exhibits three principal tendencies depending on the group: (1) a green social democratic one that emphasizes ecological modernization; (2) an anti-capitalist and radical one which believes that, ultimately, capitalism must be transcended to achieve a safe climate; and (3) an in-between one that recognizes climate justice issues but is not explicitly anti-capitalist (Baer 2012). Whereas groups that reflect the first tendency generally do not give a great deal of attention to social or environmental justice issues, those that fall into the latter two tendencies do. Individual climate activists and organizations which participate in the climate movement as a whole and its anti-coal campaigns draw upon a broad range of green philosophies in articulating their ideas about climate change and even individuals in a particular group often represent various perspectives.

Stuart Rosewarne, James Goodman, and Rebecca Pearse (2014) in dual capacities as social scientists and climate activists interviewed 25 Australian climate movement between 2008 and 2010. They assert that their sample was largely representative of the segment of the Australian climate movement that organized a series of climate camps and included 11 participants from Sydney, with the remainder from New-castle, the Hunter Valley, Brisbane, and Melbourne (Rosewarne, Goodman, and Pearse 2014: 19). Six of their climate activists belonged to environmental NGOs, three to Friends of the Earth in particular, two to socialist groups, six to a student organization, and one was a Green Member of Parliament. The climate camps attended by the climate activists were "set up in locations to expose, confront, and obstruct activities at emissions-intensive enterprises or coal facilities to challenge the politics that sanctioned energy-intensive and fossil-fuel based preoccupations" (Rosewarne, Goodman, and Pearse 2014: 118). Rosewarne, Goodman, and Pearse (2014) make a distinction between "climate pragmatists," who tend to be associated with the mainstream environmental NGOs, such as the Climate Action Network Australia, and "climate radicals," who belong either to some of the more progres-sive environmental NGOs, such as Greenpeace and Friends of the Earth, or to an array of groups ranging from Rising Tide, which has largely become defunct in Australia, to socialist organizations, to local climate actions groups (CAGs), many of which have gone into abeyance in recent years. Whereas the climate pragmatists accept the climate science, they place considerable faith in global carbon markets, renewable energy sources, and even various forms of geoengineering as offer-ing "decarbonisation within the existing status quo" (Rosewarne, Goodman, and Pearse 2013: 5). In contrast, the climate radicals seek to respond to the root causes of climate change, including overconsumption, and even capitalism, through "pro-grammes for sufficiency" (Rosewarne, Goodman, and Pearse 2014: 5). However, the interviewers found that the climate radicals are a disparate lot, ranging from those focusing on community-centered pre-figurative initiatives to those calling for some sort of transformation of economic relations, to those viewing the state as an ultimate source for a robust climate policy.

Rosewarne, Goodman, and Pearse (2014: 150) argue that the Australian climate movement had reached an impasse from its apex in 2008–2009 with the failure of the UN climate change conference in late 2009. However, over the past decade or so, Australian climate activists have drawn some cautious optimism from the UN climate change Paris Agreement in late 2015. While components of the Australian climate movement, such as the national Climate Action Summits and numerous local climate action groups, have gone into abeyance, in recent years, the Australian climate movement has seen a resurgence of sorts, particularly in the form of anti-coal and anti-coal seam campaigns. In part these campaigns have drawn from earlier campaigns but newer campaigns, particularly the Lock the Gate and Stop Adani, the student climate strikers, and Extinction have emerged.

The great majority or grass-roots groups in the climate movement belong to a green social democratic wing which urges lobbying politicians and persuading businesspeople to embrace a regulated green capitalism which would result in a reduction of greenhouse gas emissions. They favor a program of ecological modernization which would entail a strong shift in the use of renewable energy sources (particularly solar and wind power), energy efficiency, electric vehicles, and public transportation. Most of the green social democratic groups have been open to an emissions trading scheme and supported with some reservations the ALP-Greens Carbon Pricing Mechanism under Prime Minister Julia Gillard that was abolished by the conservative Coalition government under Prime Minister Tony Abbott.

Eco-socialists, which constitute a minority tendency, albeit a vocal one, within the Australian anti-coal movement, maintain that climate change mitigation ultimately requires the transcendence of capitalism, but because this cannot await its collapse, far-reaching transitional reforms are necessary for the interim, such as public ownership of mining and utilities; expansion of public transport; minimization of car use; redistributive policies; greatly reduced consumption; and a steep carbon tax at the site of production, but one which adequately compensates low-income groups. However, virtually all individuals and groups participating in the Australian climate movement favor a program of ecological modernization in which fossil fuels, particularly coal, are replaced by renewable sources of energy which Australia has in abundance. Actors in this wing include members of the Socialist Alliance, Solidarity, Victorian Socialists, and the Socialist Alternative; eco-socialists within the Green Party and perhaps even ALP; and an assortment of independent eco-socialists and eco-anarchists.

Rising Tide and the Stop T4 Campaign

Heralding the slogan "leave it in the ground" one contingent of the climate movement both internationally and in Australia sought to halt the mining of coal, the extraction of coal seam gas (CSG), the domestic use of both these fossil fuels, and their export to other countries. The International Rising Tide Network, a grass-roots collaboration among groups and individuals, was a key player in this campaign and sponsored Fossil Fuels '08 and '09 rallies in the United Kingdom, the United States, Australia, and South Africa (Cooke 2010: 425). Climate camps directed at coal mines, coal ports, and coal-fired power plants were organized in several countries.

Of the various climate action groups, Rising Tide has been the staunchest opponent of Australian coal exporting activities. Rising Tide Australia was established in 2004 in Newcastle, New South Wales. Connor, Freeman, and Higgenbotham (2009: 496) describe the "social field of coal-mine development in the Hunter Valley" as "an intensely contested arena." The opposition to coal mining is a diverse "community of interests," consisting of local residents, wealthy vignerons, horse stud associations, "ferals" or alternative people, and environmental groups (Connor, Freeman, and Higgenbotham 2009: 496). Along with Greenpeace, Rising Tide led the struggle to stop the Anvil Hill mine in October 2006. Building upon the concern of local people about environmental degradation, water shortage, pollution, and the threat to biodiversity from coal mining, a week before the 2009 Copenhagen climate talks, Rising Tide staged a day-long blockade of the rail line that feeds into the coal port at Newcastle. Additionally, for several years it organized an annual "People's Blockade of the World's Biggest Coal."

Rising Tide professed an explicit climate justice stance with a distinctive direct action emphasis. It argues: "Climate justice is a political practice founded on empowerment of marginalized communities, resistance to corporate control of global affairs, and to ending the growth addiction inherent in global warming" (quoted in Evans 2010/2011: 202). While, Rising Tide advocated a transition to a low carbon, low consumption economy, it stressed that this transition should not most impact low-income communities, socially discriminated communities, or low-waged employees of industries dependent on fossil fuels.

Its anti-capitalist perspective harbors no illusions about greening the state. Steve Phillips, a Rising Tide representative, stated at a global warming conference at the University of Sydney on March 2, 2007: "Coalmining companies have governments . . . wrapped around their finger, and anybody who dares call for a contraction of the coal industry is publicly caned as a dangerous extremist" (quoted in Evans 2010/2011: 221). The organization emphasizes the role of the culture of consumption in contributing to greenhouse gas emissions and thus climate change. For example, on April 1, 2008, it unfurled a banner in the Westfield shopping center in Newcastle that highlighted two issues: that Australian consumerism leaves Australia with a trade imbalance that is offset by expanding coal exports that prop up a non-sustainable local economy, and consumer demand for cheap goods from China and India accelerates the construction of new coal-fired power plants in those countries in order to provide the necessary energy and thus contributes to climate change (Evans 2011/2011: 213). During 2009, Rising Tide co-convened the first Climate Action Summit and jointly organized a peaceful sit-in at Parliament House a week before the Copenhagen climate conference.

Unfortunately, Rising Tide Australia went into abeyance, although much of its initial impetus had for a while been absorbed by an alliance called the Coal Terminal Action Group (CTAG) consisting of various assemblages in Newcastle and the Hunter Valley, largely spearheaded by Hunter Community Environment Centre. CTAG protested plans on the part of Port Waratah Coal Services, of which Rio Tinto and Xstrata are the controlling interests, to build a fourth coal terminal (T4), that would have the capacity to ship some up to 70 million tons of coal annually. In 2013, Port Waratah Coal Services delayed pushing forward with its proposed project for a variety of reasons, but already had received permission to build from a willing New South Wales.

Lock the Gate Alliance

In a path-breaking alliance, farmers and graziers joined with climate activists in the Lock the Gate Alliance, a national coalition of 122 community groups and hundreds of individuals alarmed by CSG (Lock the Gate n.d.). Formed in 2010, following community meetings in New South Wales and Queensland, Lock the Gate has spread to other states, where it remains active. With increasing numbers of landowners locking the gate and putting up signs telling CSG companies not to enter, Drew Hutton, the national president of Lock the Gate, sees the movement as a significant conduit between rural people and urban climate activists. The Alliance seeks to protect Australia's water systems, agricultural lands, bushlands, wetlands, and wildlife and the "health of all Australians", as well as "Australia's Aboriginal and cultural heritage" (Lock the Gate n.d.). It has a policy of engaging in "peaceful protest" which entails not damaging the "equipment, apparatus or property of others" (Lock the Gate n.d.). Furthermore, it calls upon the federal government to impose a moratorium on coal seam gas and other unconventional gas mining, end subsidies to fossil fuel companies, and take other environmental protection measures. The Alliance has not articulated a full environmental justice perspective that, unavoidably, would have to recognize how Australian farmers and graziers have a history of environmental injustice with respect to Indigenous people.

The Alliance seeks to protect Australia's water systems, agricultural land, bushlands, wetlands, and wildlife and the "health of all Australians" as well as "Australia's Aboriginal and cultural heritage" (Lock the Gate n.d.). It has a policy of engaging in "peaceful protest" which entails not damaging the "equipment, apparatus or property of others" (Lock the Gate n.d.). The Alliance urges its members to lobby politicians and to take the "Call to Country everywhere it needs to go – pin it to the doors of churches, financial institutions, chambers of commerce, and the offices of federal election campaigns" (Lock the Gate n.d.).

The Lock the Gate Alliance has collaborated with other campaigns opposing coal seam gas exploration and drilling. For example, its members participated in a march against CSG extraction in Lismore, a country town situated in north-eastern New South Wales. Lock the Gate has explicitly refrained from discussing climate change-related topics, apparently in an effort to alienate graziers who make be climate change skeptics. Its Facebook page

> specifies that people are not to post messages about renewable energy strategies, because it is outside the issues which Lock the Gate concern: "Topics like Chemtrails, renewable energy strategies, carbon tax and other things not related to the Lock the Gate Alliance campaign will be removed."
>
> *(Mercer, Rijke, and Dressler 2014: 293)*

Thus Lock the Gate at best operates at the edges of the climate movement rather than explicitly within in.

The coal seam gas issue has divided various communities in New South Wales and Queensland. Grubert and Skinner (2017) conducted research on this polarization in the country town of Gloucester in New South Wales which has a history of dairy production and logging and more recently coal mining and cattle production. When Pacific Power was granted a license to explore in the region for coal seam gas, the town experienced a struggle between a pro-coal seam gas faction and an anti-coal seam gas faction. Advance Gloucester supported coal seam gas development on the grounds that it would create jobs, both within the industry and in the larger community. Groundswell Gloucester opposed coal gas development on the grounds that it would damage tourism to the region and the sale of local agricultural products. Whereas the opponents of coal seam gas development tended to be people who were relative newcomers to Gloucester and tended to have more formal education and/or had professional experience, the pro-coal seam development faction appealed to family history in the region and/or the future economic viability of the town.

The Stop Adani campaign

If built, the Adani Carmichael coal mine in Queensland's Galilee Basin is projected to be the be the largest export coal mine in Australia, one in which thermal coal would be shipped through the Great Barrier Reef via a proposed new port at Abbott Point. The Stop Adani campaign received a jumpstart when the Mackay Conversation Group (MCG), a small environmental group based in Mackay, Queensland, temporarily halted Adani's plans to build a massive coal mine by "winning a case over the need to protect threatened species at its mine site" (Beresford 2018: 256). While the Abbott government quickly orchestrated a reversal of the MCG's victory, larger environmental groups, joined in the campaign to stop the planned Adani Carmichael mine. Later in 2015, Greenpeace launched a public email campaign targeting executives of Australia's four largest banks, to publicly rule out advising or financing Adani Mining on Carmichael coal mine and associated infrastructure project in Great Barrier Reef. This action was followed by public protest as part of Global Divestment Day organized in Australia by 350.org on February 13–14, 2015, and as part of 450 events in 60 countries to ask governments, universities, and financial and religious institutions to stop investing in fossil fuels industries. These early efforts led to the formation of the Stop Adani Alliance as a broad-based coalition of environmental NGOs and other progressive groups. By 2017, over 3 million people from Australia and other countries gave support to the Stop Adani campaign. Many environmental NGOs, nature conservation groups, Greens politicians, as well as the former ALP minister of climate change Peter Garrett have become part of the Stop Adani campaign in some capacity (Krien 2017: 22). Bob Brown, the charismatic former leader of the Australian Greens, "officially launched #StopAdani with 13 other environmental groups" on March 22, 2017 (Beresford 2018: 322).

The Stop Adani Alliance states on its website, "We stand together to stop Adani Carmichael Coal Mine, Rail and Port Project because it will fuel catastrophic climate change, and the project trashes Indigenous rights, land, water, Reef and culture." The Alliance calls for a ban on new coal mines and expansions in Australia, an end to public subsidies for fossil fuel projects, and, with regard to climate justice, solidarity with Traditional Owners and Aboriginal and Torres Strait Islanders who are standing up to protect their land and cultures. The Stop Adani Alliance includes the Bimblebox Alliance, Beyond Zero Emissions, the Bob Brown Foundation, Environment Victoria, Friends of the Earth Australia, GetUp!, Greenpeace, Jesuit Social Services, Market Forces, North Queensland Conservation Council, Sea Shepherd Australia, and many other groups. An Australia Institute (2017) survey of 1,421 Australians revealed that 44 percent of them opposed Adani's plans to build the Carmichael mine, while 30 percent supported it, with only 8 percent strongly supporting it. Furthermore, 68 percent of respondents opposed a subsidized federal government loan of up to $1 billion to Adani via the Northern Australia Infrastructure Facility. Despite the fact that hundreds of thousands of Australians have participated in Stop Adani campaign events and petitions, the Coalition and the ALP have by and large been indifferent to their concerns. While Adani initially claimed that its proposed Carmichael mine would create 10,000 jobs, more recently it has reduced this to 1,464 jobs (Brett 2020: 13).

On August 25, 2019, a federal court dismissed an appeal by the Australian Conservation Foundation opposing the federal government's approval of the Adani Carmichael coal mine project. The court maintained that the environment minister had properly considered the potential impacts of greenhouse gas emissions resulting from the Adani project upon the Great Barrier Reef, a World Heritage site. Three days earlier, the Wangan and Jagalingou people represented by tribal elder Adrian Burragubba lost a native title claim over the proposed Adani mine in an appeal against a previous Brisbane Supreme Court ruling that the granting of leases in the area was legal.

Student Strike 4 Climate Action

The year 2019 saw the entrée of two climate movements almost simultaneously that revived a struggling climate movement in Australia, namely the high school climate strikes initiated by Greta Thunberg's solo climate strike in part of the Swedish Parliament and Extinction Rebellion. School Strike 4 Climate Action held its first Australian action on November 30, 2018, when over 15,000 students walked out of their schools (Murphy 2019: 10). The student strikers conducted another strike on March 15, 2019, as part of Global Day of Action and have been calling for 100 percent renewable energy by 2030, the halt of the Adani coal mine project, and government action on climate change.

On September 20, 2019, Australian student climate strikers joined forces with student strikers around the world in the lead-up to the United Nations Climate

Action Summit in New York City on September 23, at which Swedish climate student striker Greta Thunburg lashed out at delegates' lack of inaction on climate change. In Australia, the student climate strikers at their various rallies around the country received support from trade unions, community organizations, churches, the Australian Medical Association, as well as tech company Atlassian which encouraged its 3,500 employees to join the strike (Hinman 2019). Unions that supported the student climate strike included the National Union of Workers, the National Tertiary Education Union, United Firefighters Union, Hospo Voice, the Victorian Allied Health Professionals Association, and the Nation Union of Students.

The September 20 strike took place in Sydney, Melbourne, Brisbane, Adelaide, Darwin, Perth, and smaller communities, attracting reportedly 300,000–400,000 demonstrators (Joshi 2021: 201).

Extinction Rebellion

Extinction Rebellion (XR) protests followed in the wake of the student climate strike in Australia. XR emerged initially in the UK in April with the aim to "[t]ell the Truth and act as if that truth is real on the climate and ecological emergency" and given the prospect of extinction due to the ecological crisis to "take radical collective action, which means engaging in a rebellion against the government" (Hallam 2019: 19–20). Extinction Rebellion poses three principals demands:

• Governments must tell the truth by declaring climate and ecological emergencies, and work with other social institutions in communicating the urgency for action;
• Government must act immediately to halt biodiversity loss and reduce greenhouse gas emissions to net zero by 2025;
• Governments must create and be led by the decisions of citizens' assemblies on climate and ecological justice.

XR quickly spread around the world, including Australia, where it started out with small meetings in February 2019, leading to small XR rallies in different cities. The election of Scott Morrison as the prime minister, who had replaced Malcolm Turnbull, spurred on large demonstrations around Australia. XR rebellions took place in various Australian cities, including Melbourne, Sydney, Brisbane, and Perth, in late 2019. In the case of Brisbane, on August 6, XR protesters closed down four major city interactions, and in Melbourne, on September 14 some 200 XR activists occupied Melbourne's Princes Bridge, which crosses the Yarra River between the CBD and the arts cultural center, disrupting eight Metro suburban train routes (Gregorie 2019: 9). On October 7, XR staged a protest in Melbourne as part of XR's international Week of Rebellion. XR activists formed a protest camp in Carlton Gardens, adjacent to the

Melbourne CBD, where they conducted people's assemblies and workshops on various issues, such as nonviolent direct action. Based partly upon partisan observation of an Extinction Rebellion rally in the northern inner-city Melbourne suburb of Brunswick, Samuel Alexander and Brendan Gleeson (2020: 131) propose a Rebellion Hypothesis which states "that Extinction Rebellion and related movements *will* expand, as behavioural shifts in society (or psychological tipping points) are provoked by the ongoing deterioration of Earth Systems and rising existential threats to the community of life."

Socialists in the climate movement

In the period of 2008–2010, the Democratic Socialist Party (DSP) and the Socialist Alliance acted as the most significant far-left organization in the Australian climate movement. The DSP came out of the Trotskyist tradition and formed in the late 1960s as Resistance as part of the larger anti-Vietnam War movement. In 1971, Resistance reconstituted itself as the Socialist Workers League, which evolved into the Socialist Workers Party (SWP), taking its name from the US Trotskyist Socialist Workers Party, but renamed itself the Democratic Socialist Party in the early 1990s (Percy 2005; Robinson 2019: 14–15). The original Socialist Alliance had been an umbrella group for several socialist groups. In 2001 the International Socialist Organisation (ISO) and DSP created the Socialist Alliance as a far-left wing electoral apparatus. Due to infighting within the Socialist Alliance, the ISO and other socialist groups left the umbrella group. The DSP operated the Socialist Alliance as a sort of front organization for several years, but decided to collapse itself into the Socialist Alliance in January 2010.

The Socialist Alliance publishes a newspaper *Green Left*, an important information source for the movement with its weekly articles on climate change impacts and climate politics, internationally and in Australia. In *Green Left* and at public events, such as its annual Green Left comedy debate, generally moderated by Rod Quantock, a renowned left-wing comedian, Socialist Alliance argues for eco-socialist analyses of and social justice approaches dealing with climate change. Various members have been highly active within the climate movement, however, more in the past than in recent years. The Socialist Alliance organized a conference on "Climate Change/Social Change" in Sydney on April 11–13, 2008, and "Climate Change/Social Change II" conference in Melbourne on September 30–October 3, 2011. Both conferences featured as keynote speakers John Bellamy Foster, the editor of *Monthly Review and* probably the leading eco-socialist scholar in the world, and Canadian Ian Angus, founder of the Ecosocialist International Network.

Socialist Alliance has developed a Climate Charter, which stresses that developed countries like Australia should take the lead on climate action. Periodically the party updates its Climate Charter that concludes: "Change the System, not the Climate." The Socialist Alliance argues the planet and the welfare of future generations must come before corporate profits. In what the Socialist Alliance now terms a 13-Point Climate Action Plan, it delineates the following demands:

- "Affirm Aboriginal sovereignty across the continent and seek Traditional Owners leadership of land, water, fire, and resource management to protect country";
- "Set immediate emissions cuts to targets to reduce net emissions to zero as soon as possible, including a target to achieve 100% renewable energy within a decade, and annual emissions reductions targets ramping up to a least 10% a year";
- "Bring power industries under public ownership";
- "Ban fracking: shut down and cap unconventional gas wells";
- "Redirect heavy industry, including steel and aluminium, and re-tool empty car factories, to manufacture essential products and infrastructure for the transition, such as renewable energy generators, public transport vehicles, rail lines and power transmission lines";
- "Start the transition to a zero-waste economy";
- "Retrofit energy efficiency measures in all buildings";
- "No nuclear power plants. End uranium mining";
- "Ban the logging of native forests and move to tree plantations and alternative crops for fibre and timber";
- "Support farmers to move to agroecology";
- "Expand public transport to enable all urban residents to use it for their regular commuting";
- "Include emissions from the military in the national greenhouse accounts";
- "Australia should stop blocking international treat negotiation and push instead for a binding international agreement for 90% emissions cut by 2030" (Socialist Alliance 2020).

While I agree with all of the points in the 13-Point Climate Action Plan, I could add several more, particularly on issues of social justice, the culture of consumption, and air travel.

Socialist Alliance members as well as members of Solidarity, a small socialist organization, often attend climate rallies and climate action conferences and participate in various anti-coal and anti-coal seam gas campaigns, including some of the ones discussed in the following.

In September 2020, *Green Left*, the Socialist Alliance weekly newspaper, published an "Ecosocialist Manifesto" calling "for an egalitarian cooperative road to an ecosocialist future" for discussion and development at a series of ecosocialism conferences in several cities the following month (Socialist Alliance 2020).

Solidarity has made climate change one of its campaign streams and regularly publishes articles on climate change topics in its monthly magazine *Solidarity*, many authored by Chris Breen based in Melbourne. It produced a pamphlet titled *Climate Change Is a Class Question*, which hopes that working-class industrial action – "green bans" – will in due course force polluting companies to cease production.

The Socialist Alternative has tended to operate at the fringes of the climate movement, referring to it as a "middle-class movement" rather than a "working-class movement," but some of its members have been present at climate rallies over the years. Until the Victorian state election of 2018, when it became part of the Victorian Socialists, it eschewed electoral politics, adhering to the "IS [International Socialist] tradition's belief that the task of the left was to build the Leninist party and wait for a revolutionary situation rather than dabble in electoral politics" (Robinson 2019: 127). In recent years Socialist Alternative has been more open to grappling with the ecological crisis and climate change, a move that has been manifested in its articles in its newspaper *Red Flag* and its journal *Marxist Left Review*.

In the 2018 Victorian state election Stephen Jolly (a high-profile socialist on the Yarra City Council), the Socialist Alternative, and the Socialist Alliance formed an alliance called the Victorian Socialists. In their Manifesto, in terms of addressing the ecological and climate crises, they proposed the following policies:

- "Real renewable energy targets and emissions targets";
- "Real transition fuels, not gas";
- "Public investment program for a renewable revolution";
- "Established the Great Forest National Park, end old-growth logging and revive a sustainable timber industry"'
- "Fix the recycling crisis" (Victorian Socialists 2018: 45–48).

Conclusion

I often hear climate activists in Australia say that we do not have enough time to transcend global capitalism to be able to create a safe climate for humanity. Thus, they argue that climate activists need to collaborate with more supposedly progressive corporate leaders and politicians in tackling the climate crisis within the parameters of the existing global political economy. In my view, combatting climate change and global capitalism go hand-in-hand. While the more enlightened corporate elites and their political allies may permit some measures that contribute to climate change mitigation, they will certainly not consciously permit the eventual demise of global capitalism and the emergence of an eco-socialist world system. As I have argued and tried to demonstrate in *Global Capitalism and Climate Change* (2012), green capitalism and existing climate regimes are not sufficient to mitigate climate change in any serious vein. How can we expect the system that created the problem to solve the problem?

Critiques of global capitalism both in general and in Australia tend to be muted within the contours of the anti-coal and the anti-coal seam movements. In terms of social class composition, the climate movement and anti-coal and anti-coal seam gas components appear to be comprised of four groupings: (1) middle-class, university-educated, liberals who tend to be concerned about the loss of natural beauty, iconic species, and other elements of invented Australian culture; (2) some anti-capitalist socialists and anarchists; (3) landowners and cattle and sheep farmers concerned about their own property; and (4) some Indigenous communities and groups. The groups within these movements that tend to be the most anti-capitalist in their orientation are Friends of the Earth, Rising Tide, and the Socialist Alliance. While a substantial number of Greens apparently are eco-socialists or "water melons," the leadership of the national Green Party and the various state Green parties, however, tend to exhibit a green social democratic stance toward capitalism, one which most actors, at both the individual and collective levels in the Australian climate movement and anti-coal campaign, tend to share (Fredman 2013). In Australia as elsewhere, as Patrick Bond (2012: 213) observes, in the short term eco-socialism is not on the agenda because "there is not a sufficient cadreship and network of organisations capable of connecting the dots, global or in any national setting." Nevertheless, it is imperative that eco-socialists work within climate action groups, including the ones with a strong anti-coal stance, to more explicitly identify global capitalism as the "elephant in the room" as the ultimate cause of environmental degradation and climate change.

In Australia, the recent divestment from fossil fuels campaign may provide a foothold for furthering an anti-capitalist perspective, although one fraught with many dilemmas. Go Fossil Free Australia is a relatively recent campaign which is requesting that universities (including the University of Melbourne), religious institutions, and local government councils divest their holding in "high-carbon assets." It is also asking individuals to contact their superannuation or retirement funds and banks, imploring them to do the same. Go Fossil Free Australia collaborates with Market Forces, an affiliate project of Friends of the Earth Australia and a member of the BankTrack network of organizations "working toward responsible investment in the banking sector." The group also collaborates with 350.org, a climate action group founded by Bill McKibben, which presents itself as a "grassroots global movement working to solve the climate crisis." Greenpeace now operates a "Dirty Banks campaign" against the financing of coal projects in Australia.

Transcending the devastating impact of coal and coal-seam gas production and consumption upon ecological and climatic conditions will require policies that break the existing coal-industry/state nexus, but this would require an exertion of political will that does not presently exist in the two major Australian political parties. Such policies may require shifting away from the trend toward privatization of the past 35 years or so in Australia and revisiting public ownership, nationalization, or socialization of coal mining. Presently, neither the Greens nor most of the groups, except for eco-socialists in the Greens, the Socialist Alliance, and Solidarity, in the anti-coal and anti-coal seam gas movements raise this possibility, thus ultimately constraining the notion of climate justice. Socialization or public ownership, even in the form of cooperatives, has the potential to facilitate the shift from fossil fuel energy sources (with their government subsidies) to renewable energy and make what has come to be known as a "just transition" for workers who would be displaced from the former and transitioned to the latter.

While the COVID-19 pandemic has not affected Australia as adversely as it has most countries around the world, with periodic lockdowns, it has had a dampening effect upon particularly holding of rallies on the part of climate action groups, although Extinction Rebellion began conducting small rallies in early 2021 as restrictions were eased. Nevertheless, the Climate Council and Breakthrough have been conducting online forums.

5

ENGAGING WITH THE AUSTRALIAN CLIMATE MOVEMENT

An autoethnography of a climate justice activist

Over the course of much of my career, I have defined myself as a scholar–activist in the sense that I have attempted to merge theory and social action (Baer 2021). While my initial work on climate change was theoretical, I quickly became involved as a partisan-observer of the Australian climate movement. In this chapter, I present an autoethnography of my involvement as a climate justice activist in that movement. Autoethnography is a research method which relies upon a researcher's personal experience to describe and critique cultural beliefs and social structures and entails intense self-reflection or reflexivity (Adams, Jones, and Ellis 2015).

On February 8, 2008, I began to conduct observations and became involved in the emerging climate movement in Australia while attending the Climate Movement Convergence conference at Northcote High School in Melbourne. While most speakers and workshop organizers at the conference proposed strategies of adaptation and mitigation that clearly sought to address climate change within the parameters of "green capitalism," such as writing letters to and lobbying politicians and business leaders, I found that many conferees were committed to mass action and moving beyond "business as usual." A highlight of the conference was the launching of a report titled *Climate Code Red: The Case for a Sustainability Emergency*, authored by David Spratt and Philip Sutton (2008a), long-time social activists, and sponsored by Friends of the Earth Australia. Spratt and Sutton (2008b) expanded their report into a book, which became a household item in the Australian climate movement, but also received attention beyond Australia's shores.

My involvement with the climate movement continued with attendance at the first Climate Action Summit in 2009. Over 500 registered attendees from a wide diversity of groups attended the summit, which passed a resolution demanding that Australia participate in an international effort to reduce global levels of carbon dioxide to 300 ppm no later than 2020, that Australia's CO_2 emissions should be reduced by at least 60 percent by 2020 and 90 percent by 2030 (from 1990 levels),

DOI: 10.4324/9781003202530-6

and that the government enact a policy of 100 percent renewable energy by 2020 and establish a moratorium on all new coal- and gas-fired power plants immediately, revolutionize energy efficiency, improve mass transportation, foster agricultural biological resistance, and implement a moratorium on native forest logging.

During the second half of 2009, while on a six-month study leave from the University of Melbourne, I conducted extensive participant observation on the Australian climate movement at numerous climate action events in Victoria, South Australia, and New South Wales. Since then, I have continued to be involved in climate activism in a multiplicity of ways, by being involved in various organizations seeking to address climate change and attending numerous conferences, panel discussions, and seminars touching on climate change and the ecological crisis as well as presenting seminars, lectures, and talks in both academic and community settings on these topics as well as eco-socialism. In this chapter, I provide a selected overview of my participation and observation of the Australian climate movement.

Climate rallies

In this section I describe a selection of climate rallies that I have attended since 2008 in three states, namely New South Wales, Victoria, and South Australia. The organizing of climate rallies all too often entails jockeying for position among various climate action groups. For example, in late 2011, GetUp!, a progressive multi-issues largely online organization, coordinated a series of "Say Yes" in support of the Gillard government's proposed Climate Pricing Mechanism, but stage-managed the event "to the point that other climate-related groups were asked not to distribute information" (Tietze and Humphreys 2012: 19). Having attended the rally in Melbourne, I noticed many of the attendees looked more mainstream than those who had attended earlier climate rallies, with GetUp's messaging being more mainstream, emphasizing a need for renewable energy while sidelining deeper social justice issues.

Climate camp and rally in Helensburgh, New South Wales – October 9–11, 2009

Several of the direct actions against coal have been part of the Camps for Climate Action movement, which was particularly active during 2009. The Climate Camp movement commenced when climate activists affiliated with Friends of the Earth, Rising Tide, the Australian Student Environment Network, and other groups organized a Camp for Climate Action Australia on July 10–15, 2009 in Newcastle. Protesters were reacting to expansion plans on the part of Newcastle Coal Infrastructure Group, led by BHP Billiton, for the city's coal port which had been approved by the New South Wales government. The organizers of the camp, which attracted a few thousand protesters, maintained that it was necessary to ensure a "just transition coal to clean energy – not an expansion of the coal industry" (Burgmann and Baer 2012: 298). The campers shut down the coal-port train line for an entire day. In the aftermath of the event, the campers joined local Newcastle residents and union members in a picket at the office of New South Wales Treasurer Michael Costa, calling for green jobs and opposing plans to privatize electricity provision.

Upon arriving at the climate campsite situated on a football field on the edge of Helensburgh in the morning of October 10, I first checked into the tent operated by Resistance, the youth group of the Socialist Alliance, given that I was slated to give a talk later in the day. I attended a panel discussion on the need for green jobs at which Graham Brown, a Socialist member, and John Kaye, a Greens New South Wales parliamentarian, spoke. When I asked the speakers about their thoughts about socializing energy production in Australia, Kaye noted that there is a need to reverse the trend toward privatizing energy production in the country and thanked me for my question in my conversation with him following the panel discussion. He told me that he had earned a PhD in electrical engineering at the University of California–Berkeley.

The next day, protestors assembled at the footie field where the climate camp was situated and marched to the Peabody Coal Mine situated on the edge of Helensburgh. A Socialist Alliance member told me that there had been some "red baiting" going on in the climate movement in New South Wales. Upon reaching the gate outside the coal mine site, many of the protestors staged a peaceful sit-down. I had a brief conversation with Lee Rhiannon, a Greens New South Wales parliamentarian who was running for a seat in the federal Senate, which she won. She told me that she grew up in a Communist Party family. In her address to the protesters, Rhiannon referred to the coal industry as a "jobs killer" and spoke about the need for renewable energy production. She congratulated the four protestors who had locked onto a conveyor belt in the coal mine site.

Rallies at Hazelwood coal-fired power plant in 2009 and 2010

As Chubb (2014: 39) observes, "For environmentalists, the eight tall chimneys of the Hazelwood power station are a hated symbol of the lack of progress on dealing with climate change in Australia, a source of outrage at the dominance of electricity generators." The first rally calling for the closure of Hazelwood occurred in 2005 when some 50 demonstrators unfurled a "Quit Coal" banner with 12 of them occupying the grown-coal pit and two locking themselves to coal-dredging equipment (Chubb 2014: 39).

The State Electricity Commission (SEC) of Victoria opened the Hazelwood coal-fired power plant near the Latrobe Valley town of Morwell on November 24, 1964, an operation that was projected to be shut down around 2005, given that coal-fired power plants have a life span of about 40 years (Doig 2019: 61). At its height, it produced a maximum of 1,600 megawatts of electricity and approximately 58 percent of Victoria's total electricity supply. As part of its large-scale privatization program, the Coalition government of Premier Jeff Kennett sold the Hazelwood power station and its adjacent brown coal mine to British Power in 1996 (Doig 2019: 71). In 2003 British Power created International Power, which acquired a license to operate Hazelwood until 2036, some 30 years beyond its expected lifespan. International Power also acquired the nearby Loy Yang B coal-fired power plant in 2004. Coal mine fires occurred at the Hazelwood site in 1977, 2005, and 2008.

At its peak, Hazelwood employed some 3,000 people but in time was able to operate with some 500 permanent staff and about 300 contractors (Doig 2019: 72). Hazelwood workers constituted a "labor aristocracy" in that by the 2010s, the starting salary at the plant was $120,000, with the average salary being $172,00 and managers earning up to over $300,000. Hazelwood earned the dubious distinction of being Australia's dirtiest coal-fired power plant, producing 16 million tons of carbon dioxide emissions per annum (Doig 2019: 250).

I attended the "Switch off Hazelwood, Switch off Coal" rally on September 13, 2009. Approximately 500 protesters, about half of whom had spent the previous night camping nearby, gathered for a two-kilometer march to the entrance of the power plant. The company had erected a temporary fence in front of the parking lot outside the main gate, but more than 20 people managed to scale the fence to place a "Community Decommission Order" on the plant before they were tackled by police.

I also attended the "Switch off Hazelwood, Switch on Renewable Energy" rally on October 10, 2010. The mood was no less festive, but the 2010 rally did not have a climate camp and no arrests. The numbers were smaller, with some 200 protesters being present. The police presence, up to 300 officers, including several on horseback, was overkill. The authorities erected a temporary fence to prevent protesters from knocking down or scaling the perimeter, as had happened the year before. Julia Diehm from the legal support team told protestors that the state government had passed laws in 2009 to suppress "eco-terrorists," such as those at the Hazelwood rallies. She asked protestors to monitor police behavior and report incidents of police harassment to the legal support team. On the two-kilometer walk to the plant, marchers chanted: "Coal. Don't dig it! Leave it in the ground. Get with it!" Penelope Swales sang protest songs at the gate, with comedian Rod Quantock as the MC. Greens candidate Samantha Dunn, running for the Eastern Victoria seat in the state election, read a statement for federal Greens deputy leader Christine Milne: "The Greens hear you and are calling for 100 percent renewables. . . . Coal-fired power stations must be closed, and coal companies must pay for pollution." Deb Hart from LIVE, a Melbourne climate action group, called Hazelwood a "death machine." Two "climate emergency service" workers, helped by construction workers, hoisted a five-meter-high wooden structure with "Solar Power 24/7" written on it. Protestors held aluminum-foil panels, representing an alternative solar energy plant. "MP Mr. Honorable," "Andy Union Worker," and "CEO Rachel," representing the renewable energy industry, opened a mock new energy plant.

Quit Coal was a Melbourne-based collective which campaigned against the expansion of the coal-related projects in Victoria and called for a rapid shift to renewable energy sources. The group formed in 2009 with the Switch off Hazelwood, Switch on Renewables campaign, which protested continued government support of Hazelwood Power Station in the La Trobe Valley. This campaign marshaled support from Greenpeace, Environment Victoria, the Greens, and the Australian Conservation Foundation and organized various protest rallies in 2009 and 2010. The Victorian government under the leadership of ALP Premier Daniel Andrews eventually facilitated the eventual closure of Hazelwood in November 2016. In November 2012, Quit Coal became a part of Friends of the Earth Melbourne. Six local councils and 59 organizations joined Quit Coal in demanding that the Victorian government impose a moratorium on all new coal and coal seam gas projects, a campaign which succeeded. During 2012–2014, over 25 Victorian towns declared themselves "coal and gas free" communities. Quit Coal advocated public ownership of energy production and energy democracy as an essential ingredient in transitioning to 100 percent renewable energy.

I attended the "Close Hazelwood" rally at the Victorian State Library in Melbourne on November 6, 2010. The rally was organized by Environment Victoria, which involved perhaps 300 protesters, including members of the Greens, the Socialist Alliance, the Socialist Alternative, and the Socialist Party of Australia. I distributed copies of the Socialist Alliance's Climate Charter, telling people that it was the best climate charter in Australia. David Karoly, an eminent climate scientist at the University of Melbourne, told the protesters:

It is a sunny day. We have the potential of replacing Hazelwood. I am going to talk about the climate science. Young people ask who is to blame for heat waves and droughts. Hazelwood is partly to blame for causing climate change. Hazelwood is the most polluting power station in Australia. Carbon dioxide has an impact on the atmosphere for over 1,000 years. The warming causes the melting of the Arctic ice. Closing down Hazelwood will diminish the damage. We need to close Hazelwood now. Renewable energy has created jobs in Germany. Renewables are starting to become a real thing.

During his tenure at the University of Melbourne, Karoly spoke in a wide array of public forums about the impacts of climate change on Australia and the urgent need for Australia to take climate action. For his efforts, he came under frequent anonymous attacks (Gergis 2018: 164). An Environment Victoria spokesperson told the protesters, "It is important that no Victorian is left behind as we move toward a clean energy future." He noted that EV obtained financial support from Friends of the Earth and Greenpeace but pleaded with the protesters to donate to the rally, noting, "We don't have money to pay for today. Please help." The EV spokesperson listed other organizations that had provided support for the rally, including Climate Action Moreland and Lighter Footprints. Cam Walker, the executive director of

Friends of the Earth Melbourne, told the protesters: "How far we have come and how far we have to go. We are the treadmill of a profound shift."

The ALP Gillard government announced in March 2012 that it would pay GDF Suez $383 million carbon tax compensation, a payment that ended with the abolition of the Carbon Pricing Mechanism by the Coalition Abbott government (Doig 2019: 84). Tragically, in the midst of the Victorian summer of 2013–2014, one of the state's hottest, the Hazelwood coal mine caught fire on February 9, 2014. The fire burned out of control for 48 hours, with toxic smoke and gases choking the Latrobe Valley with a population of more than 70,000 people residing within 20 kilometers of the mine (Doig 2019: 4). Short-term health problems emanating from the fire included stinging eyes, sore throats, headaches, chest pains, difficulty breathing, rashes, nausea, metallic taste in the mouth, diarrhea, vomiting, bleeding noses, gums, and bleeding eyes. While the long-term health impacts of the fire are difficult to assess, it resulted in an estimated 23 deaths. The local grass-roots group Voices of the Valley expressed disappointment with the Victorian Coalition government's Hazelwood Fire Inquiry (Doig 2019: 190).

In the political fallout, the Coalition government under Liberal premier Deniss Napthine became the first one-term government in Victorian history. Incoming ALP premier Daniel Andrews promised to reopen the Hazelwood Mine Fire Inquiry, which resulted in Engie/GDF Suez having to report annually and publicly on the progress of mine rehabilitation projects. In the end, Engie/GDF Suez was forced to shut down Hazelwood in March 2017, a closure that represented a significant step in Australia's slow but contested transition from electricity generation from coal to renewable energy. Needless to say, retrenched Hazelwood workers expressed mixed responses to the closure of their workplace, as evidenced by a sign hung up on the day of the closure, underneath rows of signed hard hats on a fence, reading "HANGING OUR HATS ON A WORKER TRANSITION", directly besides "FUCK THE GREENIES". One hard hat was scrawled with "1988 to 2017→SHUT DOWN BY GREENS AND LABOR." (Joshi 2020: 183).

The incident illustrated the importance of the creation of a transition plan to protect workers impacted by closure of a wide array for fossil fuels-based industries, ranging from coal mines to coal-fired power plants. Hazelwood's closure was actually preceded by the closure of the small Anglesea coal-fired power station and coal mine on the Surf Coast of Victoria in August 2015, an event that received little media attention.

Climate camp and rally calling for the closure of the coal-fired power plant in Port Augusta, South Australia, in September 2009

On September 24–27, 2009, several climate action groups organized the Port Augusta Climate Camp situated on the edge of the Flinders Ranges in South Australia, about 60 kilometers from the rally site near two Port Augusta coal-fired power stations. On September 26, around 50 protestors, including Climate

Emergency Action Network (CLEAN) members, students, anarchists, and others including me, assembled at the rally site, demanding closure of the two power stations. The protesters marched about two kilometers into a "protected area" but were met by around 70 police, including 12 on horseback. Eventually the modest efforts of the climate rally at the Port Augusta coal-fired power plant served as an impetus for the Port Augusta community vote to replace it with a solar thermal power station, a proposal that the South Australian government adopted in August 2017 when it announced that it would construct a concentrating solar thermal facility at Port August, slated to be completed in 2020.

Walk against warming rally in Sydney on December 12, 2009

The Nature Conservation Council of New South Wales organized the rally which occurred at Martin Place in Central Sydney. A speaker told the protesters that the Melbourne Walk Against Warming rally had attracted 50,000 people. Two Socialist Alliance members told me that they estimated a crowd of 15,000–20,000. A speaker noted that the rally had attracted social justice groups, unions, and faith-based communities. Placards in the crowd indicated the presence of members of Greenpeace, the Socialist Alliance, the Socialist Alternative, GetUp!, and Make Poverty History. A male member belonging to the Spartacists and a female member belonging to the Communist Party Australia were selling their groups' newspapers. One speaker said, "We must send a message to every politician in this land – Kevin Rudd, Tony Abbott." A representative of the firefighters told the protestors that "climate change is a reality. It is union business. This year we had bushfires at a catastrophic level. We must make the polluters pay." A female speaker told that protestors: "Political will is needed. Talk to your politicians. Write a letter to a politician and a newspaper. We need a legally binding agreement in Copenhagen."

Stop HRL Rally at Victoria Parliament House – February 1, 2012

About 300–400 protesters attended the rally. A woman affiliated with both Friends of the Earth and the Stop HRL campaign served as the MC. A female spokesperson for Environment Victoria told the protesters that

> HRL [a Chinese energy firm] is planning to build a 600 mkw plant that would purportedly be 30 percent more efficient than a plant burning brown coal. The project has the support of both the state and federal governments but the community does not support the project.

Kelvin Thompson, the ALP member of parliament from the seat of Wills, told the protesters that the ALP federal government is re-evaluating the proposed project. Adam Bandt, the deputy leader of the Greens and the MP from the federal seat of Melbourne, told the protesters that the mood in Australia has shifted to

100 percent renewables, adding: "Countries must decarbonize by 2040 at the latest. HRL is a threat to the environment and economically unviable. It makes no sense to have one foot on the brake and the other on the accelerator at the same time." He said that he would introduce a bill in Parliament that would provide public funding for renewable energy. Other speakers at the rally included spokesperson for Greenpeace and the Australian Youth Climate Coalition (AYCC).

I attended an anti-HRL follow-up meeting at the Friends of the Earth Melbourne headquarters on May 21, 2012, attended by some 40 people, most of whom were young women, some were young men, and some were middle-aged men, making me the oldest person present. A male Friends of Earth spokesperson spoke about plans to door knock and letterbox in some key electorates to gain community support opposing the HRL project. He noted that "we are close to winning on HRL and have been working closely on the campaign with Environment Victoria and Greenpeace."

Climate change conferences and public meetings

In this section I describe a selection of climate change conferences and public meetings that I have attended as a climate justice activist and scholar of the Australian climate movement. These include the Green New Deal conference sponsored by the Australian Greens, the T10 launch at Melbourne Town Hall, Bill McKibben's address at Environment Victoria, Climate Summit 2011, the NTEU "Pushing the boundaries" conference, a Greens public meeting on climate jobs and the Green Bans, a Greens public meeting at Victorian Trades Hall, a Greens community forum on the Carbon Pricing Mechanism, and a Socialist Alliance forum on the Paris Agreement.

The Green New Deal Conference, October 23–25, 2009

The Green New Deal Conference sponsored by the Australian Greens was preceded by an address titled "The greening of politics" by Bob Brown, the party leader and a senator from Tasmania, at "The Spot," the building housing the Faculty of Economics and Business at the University of Melbourne in the evening of October 23, 2009. In his address, he observed:

> The climate is in a grim state. Our government accepts a carbon market. Scientists are telling us we have less than a decade to turn things around. Things are getting worse than we once thought. There was a rally outside Parliament to save solar system.

Brown showed the audience a short film about the travesties of logging in Tasmania in which he is the narrator. He lamented that the eagles in Tasmania are headed for extinction. Brown noted that the "greening of politics has been going on for three or four decades" and that the Tasmanian Greens were the world's first Greens party, adding that the German Greens leader had served as an inspiration to Australian Greens. Brown observed that "we know about the existence of 400 planets, but Earth is the only one habitable for life as we know it." In referring to "this beautiful biosphere," he added:

> This planet is in its 6th stage of massive extinction. We are destroying this planet, but we can fix it. Our Prime Minister [referring to Kevin Rudd] looks at our resource extraction and wants it to get bigger. Bligh [the Queensland ALP premier] wants to make Queensland's coal exports bigger. I have people ask me all the time how do you keep so optimistic. We are in the process of turning that around. Ralph Nader said that information is the goal of democracy. It is in our hands to turn things around. There is a problem with capitalist society. Democracy needs to be managed. Money and economics need to be the servants of humanity. Both of our parties [the Coalition and the ALP] accept growth. We need prosperity without growth. We cannot continue with the growth model. They are under the thumbs of the coal barons and polluters.

The opening plenary presentation on Saturday morning entailed a teleconferenced presentation by Tim Jackson (2009), the author of *Prosperity Without Growth*. Prior to the plenary, I had a conversation with Christine Milne, the deputy leader of the Greens and also a senator from Tasmania. She invited me to follow her into the plenary event. When I showed her a copy of *Global Warming and the Political Ecology of Health*, a book which I had co-authored with Merrill Singer, she put into her pursue, although I had intended merely to show to her, not give it to her (Baer and Singer 2009). I prudently opted not to ask for my book, thinking that perhaps her reading might do some good. Milne took it out of her purse and asked me to sign it, noting that social science is important in order to understand people's responses to climate change. When I asked her view on socializing or nationalizing energy companies, thus opening the door to decarbonizing them, she opposed my proposition. When I told a Greens member from South Australia her response to my question, he noted, "Christine is conservative." In his televised address titled "Prosperity without growth," Jackson observed:

> Economic growth is unsustainable as we have it. De-growth is unstable. The engine of growth is Economics 101. We don't have a macroeconomics of instability. The economy is about consumption and investment. Investment is the key to a new economy – an ecological investment. The public sector can accept a lower rate of return. Jobs matter.

Bob Brown fielded questions from the audience directed at Jackson.

The second plenary session was titled "Green New Deal: What Needs to Change" and featured a panel discussion facilitated by Amanda Lowey with Cheryl Buchanan, an Aboriginal woman, and Iain McGill, an electrical engineer from the University of New South Wales, serving as the panelists. Buchanan told the audience that Aboriginal people are resilient and have been subjected to genocide, water hole poisonings, rape, having their children stolen, and having competing value systems, including capitalist ventures, imposed upon them. She continued: "Aboriginal people have heard about black, red, yellow, and green new deals. Roosevelt's New Deal was an effort to rejuvenate American capitalism."

The first plenary session on Sunday morning was titled "Green New Deal – How Does Change Come About?" Steven Keen identified himself as an "anti-economist" and asserted that "economics has no concept of time." Joan Staple (visiting fellow, University of New South Wales) presented a tripartite model of democracy consisting of business, government, and the community with NGOs being a component of the latter. She stated that the new climate action groups "have warmed my heart." Although she had worked with the Australian Conservation Foundation in the past, she did not agree with their support for the CPRS, which the Greens had rejected as an effective climate change mitigation strategy. Jake Wishart, a young union organizer, noted that he had worked with Al Gore's Climate Reality Project. Christine Milne, the deputy leader of the Greens, applauded the climate movement for its efforts and said that Green parties would be collaborating with environmental NGOs at the forthcoming UN FCCC conference in Copenhagen. I managed to get the first question in during the Q&A in which I said:

> I applaud Joan's enthusiasm about the climate movement and Christine and Bob Brown's presence at the Climate Action Summit in Canberra earlier this year. I do have trouble with Joan's tripartite model of democracy. Power is concentrated. Noam Chomsky says of the US that it is a one party system – a business party with two factions – and I suspect that same is true of Australia. In the modern world, multinational corporations make and break governments and politicians. The Greens are an exception to that. What are your thoughts on the concentration of power?

Some members of the audience applauded my question. Joan replied, "Good question. I am aware of the concentration of power" and Keen agreed that I had posed a "good question."

Several New Zealand parliamentarians served on the panel titled "A Green New Deal for Aoteroa/New Zealand." The panelists noted that unemployment in New Zealand in June 2009 had been 6 percent, the GDP had declined, and water problems were anticipated for the Canterbury Plains on South Island. When I asked about what was the New Zealand Greens stance on public ownership or nationalization, one of the parliamentarians noted that is a taboo topic in his country. A series of workshops occupied Sunday afternoon of the conference, a few of which I attended, but none of which I found particularly illuminating.

After attending several other plenaries and workshops on the Green New Deal, I decided that by and large, the Australian Greens were too soft on capitalism for me to join their party. However, I did give talks on eco-socialism to both the Yarra Greens and the Darebin Greens in subsequent years and have maintained a dialogue with Greens about climate change and eco-socialism over the years.

T10 Launch at Melbourne Town Hall on February 10, 2010

The Transition Decade campaign (T10) was committed during 2010–2020 to create a "safe climate outcome" by reducing the earth's temperature between 0.3 and 0.8°C and reduce CO_2 in the atmosphere to between 280 and 320 ppm. The first two years of T10 were to focus on encouraging whole communities to say "yes" to this objective, the subsequent eight years to obtain government commitment to the plan, and the last few years to the adoption of a "climate emergency mode" during which society gears up to building a climate-safe economy. Some 700 people attended the launch of T10 sponsored by an alliance consisting of the Climate Emergency Network, Beyond Zero Emissions, and Friends of the Earth Melbourne. Following a welcome to country by a female Aboriginal elder, the event commenced with a video discussing how responding to the climate emergency required the management of human activities contributing to climate change, the adoption of a war economy framework for addressing the climate emergency, the rapid adoption of renewable energy, and a movement that includes the whole community. David de Krestser, the governor-general of Victoria, presented a speech titled a "Vision for a Sustainable World," in which he advocated shifting to more sustainable lifestyles in which people tread more lightly on the planet over the course of the next four decades for the sake of future generations, including his six grandchildren. Will Steffin, an Earth scientist based at the Australian National University, stated that the past decade has been the warmest one that humanity has experienced, observing that it will be difficult to stay under two degrees of warming. Cam Walker, the coordinator of Friends of the Earth Melbourne, observed:

> The first stage is to get the broader community to say yes [to climate action]. The second stage is to get the government on board and to build the political will to take climate action in the next four years. We need to live within ecological limits. The two major parties are into growth. We have to stop consuming so much. We need to get into yes. The next decade will entail a lot of work. I see a lot of burnout in the climate movement. Mitigation and adaptation are both important.

Christine Milne, the deputy leader of the Australian Greens and a Senator from Tasmania, said:

> We need to reconnect with the planet. It is important to build momentum. Give yourself a big round of applause. We need systemic change throughout the political system. It is amazing the tug of war between what is and what can be. The vested corporate interests and labor both believe in growth.

Twelve coal-fired power plants exist in Australia. You must get involved in politics. . . . We need to get people out of their cars and into trams, buses, and bikes.

Philip Sutton, the co-author of *Climate Code Red* and the convener of the Climate Emergency Network, stated: "We have to rebuild the economy very fast. We have to get into emergency mode and get the corporations on board." In my discussion with Chris Breen, a Solidarity member and a member of the Climate Emergency Network Facilitation Committee, we both agreed that Sutton's expectation of expecting significant climate action from the corporate sector was politically naïve given that, we as fellow eco-socialists viewed the corporate sector as constituting the driving force of the climate emergency.

Bill McKibben's address at Environment Victoria, May 26, 2010

Bill McKibben is a renowned American environmentalist and the Shumann distinguished scholar at Middlebury College in Vermont. His first book *The End of Nature*, which was serialized in *The New Yorker*, propelled him to fame as an important environmental writer, in large part because it was regarded to be the first book about climate change targeted at a general audience (McKibben 1989). McKibben has since then authored numerous books. His most recent book *Falter* draws upon his experience in building 350.org, the first global grassroots climate movement that seeks to grapple with a grim moment in which humanity is at a crossroads where it has to confront the climate crisis head-on or see civilization slip away (McKibben 2019a).

Some 125–150 people attended the event at Environment Victoria's headquarters in the Green Building on Leicester Street in Melbourne. In his informal address to the audience, McKibben noted:

> I am sorry to have to come and go. My carbon footprint rivals that of Hazelwood. We had the warmest April on record. Change has come far more quickly than we expected. The ocean is 30 percent more acidic, the planet is one degree warmer, another degree is in the works. Northern Europe has had major flooding. We have 390 ppm [carbon dioxide] outside. Politics is equally bleak around the world. Australia is one of the few places that elected a prime minister on climate change. We see how much good that has done. 117 nations at Copenhagen signed onto a 350 target. We are going to need more resilient communities in a resilient world. It is good to see campaigns on Hazelwood like this one. Power plants in DC [District of Columbia] have converted to natural gas. It is a start but we need to go further. We need a global price on carbon. There a coalition in DC between climate activists and people who live near coal-fired power plants. Much weighs on the US, China which I just visited, and Australia too.

In the Q&A following McKibben's remarks, I stated,

> There is a tendency in the climate movement in the developed world to focus on ecological modernization, renewable energy and energy efficiency. We can have solar panels on McMansions. We need to address a global economy with its treadmill of production and consumption heavily reliant on fossil fuels. We need to address that fact that many in the developed world are living high on the hog. What are your thoughts on that?

McKibben replied, "You are right. Social inequities are an issue."

In 2019, I had an opportunity to dialogue with McKibben on the state of the climate movement along with four other commentators in a *Great Transition Initiative* forum (McKibben 2019b; Baer 2019). As a long-time veteran in the global climate movement, McKibben asserts:

> We've had more successes than I imagined we would. . . , but we have yet to turn the tide: the power of people is not yet mobilized in sufficient strength to outweigh the financial majesty of the fossil fuel industry, and we continue down an ever-hotter path. Also, the price of even that mobilization has been enormous: in some parts of the world, environmental advocates are routinely murdered, and even in places where they operate with more freedom, the stress and strain are very real. I know so many people who have given over the prime of their lives to this fight. Some have been to jail, wrecked their careers, and burned out their emotional cores. They've been sued and surveilled by oil companies, attacked by guard dogs.
>
> *(McKibben 2019a: 191–192)*

Climate Summit 2011 on April 11–12, 2011

The Climate Summit 2011 was an abbreviated version of the 2009 and 2010 Climate Summits in Canberra. The event took place at the University of Melbourne and consisted of several plenary sessions and workshops on an array of topics, including the pros and cons of the proposed Carbon Pricing Mechanism; various forms of carbon pricing; and a Climate Scorecard rating the climate policies of the Coalition, the ALP, and the Greens. In a teleconferenced presentation in a panel on "A new brand of environmental radicalism," Clive Hamilton (2001, 2007, 2010), the author of three climate change-related books, asserted that Australian environmentalism has lost its way and criticized some environmental groups for supporting incremental measures. He argued that Australia needs a new environmental radicalism, not one that is concerned about turning off voters with its positions. In a panel the next day on "Carbon pricing and national politics," Adam Bandt spoke of the need to transition to a "zero emissions economy" and alluded to the flack that the Greens had caught due to their opposition to the CPRS. In another panel,

David Spratt, the co-author of *Climate Code Red*, said that he supported the CPM for pragmatic reasons (Spratt and Sutton 2008a, 2008b).

National tertiary education union "Pushing the Boundaries" Conference on April 28–29, 2011

In the opening session, Jeannie Rea, the NTEU president, told the audience:

> We have to change the way we think. A carbon tax will cost jobs. Many more men than women are involved in the climate change discourse. Men are doing most of the talking and the women are doing most of the work. Women do participate in climate rallies. Let us push the boundaries on climate change.

Ian Lowe, an NTEU stalwart and the president of the Australian Conservation Foundation, told the audience:

> The 1980s had been the warmest decade on record. The 1990s were even warmer and the period of 2000–2010 was yet even warmer. Australia is the continent with the most variable rainfall. The Holocene has been a period of climate stability. The temperature is 0.7 degrees warmer than the average of the last 10,000 years. Global emissions need to peak within ten years. The denial industry consists of conspiracy theorists, dubious experts, the cheery-picking of data, ideological fallacies, and uncertainty. Alan Jones [a renowned Sydney-based radio jock shock host] got an extraordinary hearing in Sydney.

In the second morning session, Stuart Rosewarne, a political economist based at the University of Sydney, told the audience, referring to the proposed Carbon Pollution Reduction Scheme:

> The price on carbon is completely politically pragmatic. Treasury is filled with neo-classical economists who keep the price low, $20 per ton. Greg Combet [the ALP Climate Change Minister] is a trained economist who studied political economy at Sydney Uni. The level of compensation to corporations is going to eat into the budget and reduce government revenue to invest in renewable energy. CCS [carbon capture and sequestration] is not commercially viable. Let us get away from a carbon price. Who owns coal and natural gas? We do. We need to decommission coal and stop exporting coal. We need to challenge growth. We need to refocus on labor and the work/life balance. The ALP government gave everything to the mining industry.

In the first afternoon plenary panel discussion, Tony Maher, the president of the CFMEU, said that there is progress at the UN COPs and the "world is moving to a patchwork of national climate schemes." He said that environmental groups are disorganized and engage in catastrophism. Maher went onto argue:

> A global economy with lower emissions will still have exploitation and face the issue of finite resources. The unions have seen climate change as an issue for junior officials but have not been sufficiently involved. The UN process is becoming more creative. The 33 nations with an ETS have not lost their steel industries. The Greens are going through growing pains.

Adam Bandt said: "The ALP chose the Coalition as its partner on the CPRS. We will need a huge range of policies to challenge climate change. We are going to need public investment in the grid." He discussed the negotiations between the Greens and the ALP government on developing a Carbon Pricing Mechanism. In my question during question and answer (Q&A) session, I posed my question to Bandt:

> Adam, I have a double-pronged question aside of whether the carbon price will be $20, $30, or $140. After three to five years, the CPM will transition from being a carbon tax to an ETS. Is there a single ETS that has significantly reduced emissions? Can the planet sustain this or do we need to think beyond the present global economy, replacing it with one committed to social parity and environmental sustainability?

Bandt claimed that the EU and the UK ETSs had been somewhat effective and noted that one of his aides was examining the efficacy of ETSs and he would send the findings to me, which never happened, but in retrospect, I should have sent him a reminder.

In the morning plenary the next morning, Ged Kearney, the president of the Australian Coalition of Trade Unions (ACTU), called for a transition to a "clean energy economy," adding that a "price on carbon would result in economic restructuring. We are pushing for sustainable jobs." In my question to her, I asked:

> Is the ACTU addressing the fact that we have a government that is promoting the opening up of new coal-fired power plants and the expansion of coal ports? These are policies of a government which claims to be committed to climate change mitigation.

Kearney initially danced around my question and reiterated that a carbon price would stimulate renewables. Later on during the Q&A, she noted: "Forty-seven per cent of ACTU members voted for John Howard in 2004. It would be better if we didn't have market mechanisms. We would be better off if we just had a tax." Colin Long, the Victorian NETU Division State Secretary, noting that he is a Greens member, chimed in to the discussion, arguing:

> Economic growth is fundamentally capitalist. The notion of growth with decoupling emissions by moving to decarbonization lets us off the hook. We

need to challenge growth. The trade union movement is complicit with the growth paradigm. Unions need to redefine the notion of work. Strategies to challenging the ideology of growth include shorter work hours, better work/life balance, greater workplace democracy, public sector investment in utilities. Higher-income people have higher incomes.

John Rafferty, a NTEU member based at Charles Sturt University–Albany Campus, argued that universities engage in a lot of "green sheen" or "green wash" in their sustainability programs, adding, "It would be nice to have a new world order" and that "instances of green shift in Australian universities are minimal."

Public meeting – "climate jobs and green bans – What attitude should unions have to a new coal-fired power plant & HRL" at Victorian Trades Hall on August 3, 2011

Speakers at the event included Jack Munday (leader of the Builders' Labourers Federation's famous Green Bans in Sydney during the 1970s), Steve Dodd (assistant secretary of Gippsland Trades and Labour Council), and Luke van der Meulen (president of the District Branch, Mining and Energy Division, CFMEU). Munday said: "The Green Ban movement indicated that working-class people can be environmentalists. I was a Communist Party member back then but now belong to the Greens. Global warming is the biggest political issue of all. Capitalism and socialism failed in the 20th century." Dodd observed: "Privatization resulted in the loss of 10,000 jobs in the Latrobe Valley." He argued that "Gillard's carbon tax is a catalyst for change" and did not think that the HRL project would go ahead." However, Dodd, speaking on behalf of workers in the four coal-fired power plants in the Latrobe Valley, observed: "We are not climate deniers but we do want to look at other energy producers, maybe gas. We should develop new technologies that use cleaner coal." Van der Meulen argued: "We need to do something about the way we produce energy. A transition from coal to renewables is important. Twentieth-century socialism was an environmental disaster."

Greens public meeting at Victorian Trades Hall – "Australia's clean energy future: How the Greens negotiated a deal" with Bob Brown and Adam Bandt – August 11, 2011

Brown joked, "I wore my best suit which is made of Australian wool. It makes me look conservative." In a more serious vein, he went on to note:

The Carbon Pricing Mechanism has gotten huge coverage around the world. Climate change is relentless in terms of its impact on the planet and the psyche. Adam attended all the sub-cabinet meetings. Consumerism has taken over the human psyche. The planet hit seven billion people on 30 October 2011. A climate scientist in a Hobart meeting said we can expect a 4–7 meter-sea level rise by 2100. Clearly, we cannot continue the way we have been. Environmental issues are paramount. Twentieth-century socialism was an environmental disaster. We have to take first steps [referring to the Carbon Pricing Mechanism].

Greens community forum: the politics of pollution, Brunswick Town Hall – August 29, 2015

The Brunswick Town Hall was filled to capacity, with somewhere between 200 and 300 people in the audience. The speakers at the event were Richard Dennis, director of the Australia Institute; David Karoly, an eminent climate scientist based at the University of Melbourne; and Adam Bandt, the Greens MP for the seat of Melbourne. The event occurred in the build-up to the passage of the Carbon Pricing Mechanism. Dennis reported that Blue Scopes is campaigning against a carbon price. He stated: "We need to reduce greenhouse gas emissions rapidly. We have to reduce the enormous subsidies to the polluters. Those $50 billion in subsidies need to be reinvested in renewable energy. It would take $50 billion to build new subways." Karoly spoke about the potential impact of climate change on Australia if emissions are not rapidly reduced, a pitch that he had made to numerous university and community audiences. Bandt noted that he has worked in public interest law and referring to the proposed Carbon Pricing Mechanism, stated:

> What is in the package? It is not a package that we would written ourselves. There is no reason that Australia's can't lead the world in renewables. We need to make the 5,000 biggest payers pay. One half of the money will go to compensation, with low-income earners to be overly compensated. There will be an establishment of a Climate Change Authority as an independent body. The UK has a Conservative prime minister who said emissions would be reduced by half by 2025. The UK has an independent climate change authority. We negotiated for a 60–80 percent reduction in emissions by 2050.

During the Q&A period, I was the first member of the audience who went to the mike situated in the front to pose a question. I said:

> I would be the first to argue that there is a need for clean energy in Australia and elsewhere. However, do we need as much energy as we think we need in Australia? What do the panelists think about redistributing energy to the poor of the world away from high-energy users? Is there a need to shift from the present global economy to an alternative global economy or can the present system merely be tweaked? Is there a need to nationalize or socialize the energy sector?

Denniss admitted that Australia is using more than its share of resources and referred to large houses in the country, adding that "there is a need for redistribution of income and energy. It is an important question." Neither Karoly or Bandt opted to address my question. In fact, Bandt fidgeted around and called for the next question.

Later on during the Q&A, Bandt alluded to the question of how much energy Australians use and admitted that "we need to discuss climate change from a social justice perspective." The following week when I told Karoly at an NTEU climate change workshop that Bandt had not answered my question at the Greens town hall meeting, he replied, "You will have to ask him yourself. He is a politician." Ironically, I had the opportunity to do exactly this shortly thereafter, at yet another climate change workshop at the Royal Melbourne Institute of Technology where he made introductory remarks. Before doing so, I approached Bandt, saying, "Adam, you did not answer my question at the Greens town hall meeting." He replied, "I thought that Richard Denniss did such a job answering your question that I had nothing else to say." I disagreed, noting "that there was heaps more that could have been said."

Socialist alliance forum on "Paris and after: which way forward for the climate movement? on February 17, 2016

The speakers for the forum were John Englart (Sustainable Faulkner), David Spratt (co-author of *Climate Code Red* and a key player in Breakthrough), and Andrea Bunting (Socialist Alliance and Climate Action Moreland). Englart described the UN COP25 Paris Agreement given that 195 countries had signed on. He noted that fulfillment of voluntary pledges to reduce national emissions would take the world to a 2.7–3.5-degree world, not a 1.5–2-degree world by 2050 to which the Agreement aspired. Englart added that the Agreement would come into effect after it had been ratified by 55 countries and would start in 2020. He said that the Agreement made no mention of the need to decarbonize the global economy and the emissions contributions made by airplane flights and shipping. Englart reported that Australia's emissions increased 1 percent in 2015 and electricity emissions increased 3 percent. Spratt lamented:

> The climate movement is dominated by large organizations which are not democratic and are run by boards and driven by grants and money. The recent climate march included unions and mosques. Melbourne had 150,000 at its march, the largest in the world. We need to keep building a culture of cooperation.

Bunting argued: "We have to focus on successes and on 1.5 degrees. Naomi Klein used the C word and wrote about the need for government intervention. How do we work with other groups?" During the Q&A, Spratt said: "Eco-socialists need to form a bloc within the climate movement," adding that "some socialist groups do not see climate change as a problem."

Efforts to update the climate emergency framework: From Australia to the world and back to Australia

The climate emergency framework, which started in Australia around 2008, has been adopted in many countries, particularly the US and UK. In terms of the Australian scenario, in my dual roles as an anthropologist and climate activist, I witnessed the initial development of the climate emergency framework and more recent efforts to update it in Australia, particularly at the National Climate Emergency Summit in Melbourne on February 14–15, 2020. From my perspective as an eco-socialist, I argue that the climate emergency framework seeks to operate within the parameters of global capitalism and in doing so downplays social justice issues. There is a need for the climate emergency movement to become part and part of a larger climate justice movement, not simply a climate movement that emphasizes techno-fixes, and that says, "system change, not climate change."

Historically there has been a tendency for Australia to be on the receiving end of developments that begin in Great Britain or the United States. However, in the case of one variant of the international climate movement, it appears that the climate emergency framework originated in Australia around 2008. In that year, two long-time Australian activists, David Spratt and Philip Sutton, proposed a framework for climate change campaigning based on the argument that climate change mitigation is so urgent and immense that it requires emergency measures like those adopted by the Allied powers during World War II. *Climate Code Red* was published initially in loose spiral binding by Friends of the Earth Australia with the subtitle "The Case for a Sustainability Emergency" and then republished in expanded book form later in 2008 with the subtitle "The Case for Emergency Action" (Spratt and Sutton 2008). In the same way that "code red" signals to hospital workers that a patient needs advanced life-support, Spratt and Sutton argue that climate change sends a strong warning that societies and their leaders must declare a "climate emergency," which transcends "business and politics as usual" and indicates the political will to prevent a global climate catastrophe. To create a safe climate, they contend that humanity needs to reduce the current global temperature to less than 0.5°C above the pre-industrial level and greenhouse gases to less than 320 ppm CO_2e, a highly ambitious target given present realities. Spratt and Sutton maintain that the United States can be harnessed to create a safe climate along with other countries, including Australia. In terms of climate change mitigation, they argue that Australian federal and state governments have been beholden to the fossil fuel industry and have done almost nothing to challenge it. Spratt and Sutton are deeply critical of neoliberalism, under which the corporate agenda dictates politics, ensuring that capitalism is environmentally unsustainable. They envisage a green social democracy, a highly regulated capitalism, because governments, they maintain, have a crucial role to play in planning, coordinating, and implementing the transition to a safe climate. Spratt and Sutton insist that some business people are open to concerted climate change mitigation and are part of the solution because a "safe-climate economy" that cuts emissions to nearly zero will require new technologies and products.

The Climate Emergency Network emerged as a Victorian-based endeavor that emerged as a spin-off from Spratt and Sutton's *Climate Code Red* book, which for several years served as the Bible of the Australian climate movement. Describing itself as a network of community organizations and individuals campaigning for an urgent response to a climate and sustainability emergency, CEN's objectives included: working toward 100 percent renewable energy in Australia by 2020, no new coal-fired power plants, protecting and strengthening the capacity of the earth's natural carbon sinks, and supporting carbon-neutral technologies. It provided start-up financial support to Beyond Zero Emissions, a think tank, promoting renewable energy technologies and still very active in its own right. No socialist groups belonged to CEN, but individuals belonging to the Socialist Alliance, including myself, and Solidarity were active for a while in the organization. In May 2010, I submitted a resolution to make a commitment to promote climate justice both in Australia and abroad, which CEN accepted. Unfortunately, CEN ran out of steam a few years later and became defunct for a variety of complex reasons.

The first major step in promoting the climate emergency framework in Australia commenced under the banner of Transition Decade campaign (T10), which aimed in the 2010–2020 decade to create a safe climate, reduce the earth's temperature between 0.3 and 0.8°C, and reduce CO_2 in the atmosphere to between 280 and 320 ppm. The first two years of T10 focused on encouraging whole communities to say "yes" to this objective, the subsequent eight years to obtain government commitment to the plan, and the last few years to adoption of an "emergency transition mode," during which society geared up for physical reconstruction, and promoted this process on a massive scale. The original concept of T10 was developed by Luke Taylor of the Melbourne-based Sustainable Living Foundation. SLF members developed the idea during 2009 and involved other organizations, particularly Beyond Zero Emissions, the Climate Emergency Network, and Friends of the Earth. On February 14, 2010, T10 was launched publicly to some 700 people at Melbourne Town Hall and another 500 around Australia in small venues. However, the T10 initiative also fizzled out after a few years.

The climate emergency framework started to go international, in part due to the efforts of Jorgen Randers, a Norwegian and one of the authors of the famous *Limits to Growth* (Meadows et al. 1972) paradigm sponsored by the Club of Rome, and Paul Gilding, an Australian and a former CEO of Greenpeace International. In 2010, they argued that the climate emergency plan was likely to emerge prior to 2010 when global society finally more fully recognized the threat of climate change to humanity (Randers and Gilding 2010). Randers and Gilding envisioned three phases in their plan:

- A climate war (years 1–5) which would launch a global society to reduce greenhouse gas emissions by 50 percent within five years;

- A climate neutrality phase (years 5–20) which would lock in the 50 percent emergency reductions, and move the world to net-zero climate emissions during this phase;
- A climate recovery phase (years 20–100) which would entail stabilization of the global climate system and the creation of a sustainable global economy.

Their climate war model is extremely detailed, but for purposes of illustration, I list several of their recommendations:

- Reduce deforestation and other logging by 50 percent;
- Shut down 1,000 coal power plants within five years;
- Ration electricity;
- Construct a wind turbine or solar plant in every town;
- Create huge wind and solar farms in desert areas;
- Ration use of highly polluting cars to reduce transport emissions by 50 percent;
- Gradually reduce the world's aircraft by 50 percent;
- Launch shop less, live more campaigns.

Randers and Gilding propose the creation of a climate war command on the part of countries participating in the plan, which would draw upon advice from the International Monetary Fund and the Intergovernmental Panel on Climate Change as well as various multi-national military commands, presumably such as NATO. This command structure would introduce a global tax of US$100 per ton of CO_2 emissions, which would have two aims: to fund the war effort and to alleviate the resulting hardship, particularly the poor around the world.

Due to contacts with Breakthrough, Margaret Salamon Klein, a New York-based psychologist, spearheaded the establishment of The Climate Mobilization Project (TCM) in 2015. It encourages people to adopt the climate emergency framework, particularly in the United States. TCM's advisory board includes an impressive list of actors, including Betsy Taylor, former International Representative for the Nuclear Weapons Freeze Campaign; David Spratt of *Climate Code Red* fame; Michael Mann, a world-renowned climate scientist based at Penn State University and the author of the "hockey-stick" hypothesis modeling rising global temperatures; Gus Speth, a world-renowned environmentalist, the founder of the World Resources Institute, and the co-founder of the Next System Project; Richard Heinberg, a senior fellow of the Post Carbon Institute; and Aileen Getty, the founder of the Aileen Getty Foundation, which supports organizations and individuals committed to responding to the climate emergency.

The climate emergency framework which originated in Australia has achieved some modicum of success in that it has gone viral around the world. The climate emergency framework has been adopted by various prominent climate scientists, including James Hansen, Hans Schellnhuber, and Michael Mann; Pope Francis; and world-renowned climate activist Bill McKibben. Several thousand tertiary institutions around the world have declared climate emergencies. The City of Darebin, a northern inner-city suburb of Melbourne, was the first government of any sort to officially declare a climate emergency in December 2016, spurred on by the efforts of Climate Action Darebin in which Philip Sutton is an important player. Salamon (2020: 91) observes: "In November 2017, TCM's local chapter in Hoboken, New Jersey, persuaded the city government to become the first U.S. city to declare a climate emergency resolution." Largely prompted by pressure from Extinction Rebellion, the UK became on May 1, 2019, the first national government to declare a climate emergency. On November 28, 2019, the European Parliament declared a climate emergency, and numerous national governments, including Ireland, Portugal, France, Spain, Austria, Bangladesh, and the Maldives, have declared climate emergencies, as have state and local governments in numerous countries.

The 2020 National Climate Emergency Summit in Melbourne

Back in Australia, the 2020 National Climate Emergency Summit met on February 14–15, 2020, at Melbourne Town Hall and featured over 100 speakers and some 800–1,000 attendees on each of the days. The underlying agenda was to get more Australian governments in addition to some 87 local governments, starting with Darebin and including Melbourne and Sydney, to declare climate emergencies and to pursue climate action, particularly in the form of shifting from reliance on fossil fuels to renewable energy sources. The cast of moderators and presenters constituted largely a cavalcade of stars, starring out with opening addresses by Sally Capp, the lord mayor of Melbourne, and Adam Bandt, the federal member of parliament from Melbourne and the new Greens leader. While the event had numerous sponsors and partners, including the City of Melbourne and RMIT, its main drivers were Breakthrough and the Sustainable Living Festival.

In his remarks, Bandt observed: "Barnaby and his Nats [National Party] want to build more coal-fired plants and the Labor [Australian Labor Party] opposition continues to support coal." He argued: "We face an existential crisis. A four-degree world would support a billion people, even less. Government and industry worked together in World War II. Nothing will stop the clean energy revolution" and added he would introduce a Climate Emergency Declaration into Parliament. Bandt continued: "We want to move Australia into a Green New Deal and make it an energy superpower and address social inequality" In the plenary titled "New Climate Reality Check," US climate scientist Michael Mann argued that an emphasis on climate adaptation rather than climate mitigation "avoids talking about real solutions" and alluded to the need for systemic changes, although he did not spell out what they might be. In a brief conversation that I had with him, Mann admitted that the level of consumption fostered by global capitalism is a problem, but added that is a difficult issue to address. David Spratt argued that despite the intensity of the climate crisis, the world has the economic resources to solve it with appropriate actions.

In the following plenary on "Stand Up on Leadership," Peter Garrett, a former Australian Labor Party climate change shadow minister and minister of other portfolios for six years, insisted that a "national government can be an instrument for change." Zali Steggall, an independent MP from Warringah who displaced Tony Abbott, a former Coalition prime minister who once referred to climate change as "crap" and touted coal as a gift from on high to humanity, said that she would be introducing a private member's bill to Parliament to act on climate change. Jean Hinchliffe, a student climate activist from Sydney who attended the 2019 UN climate change conference in New York, and Paul Gilding, a long-time proponent of the war mobilization framework on climate action, rounded out the second plenary.

Of all the plenaries, I found the one on "Justice and Rights" on the second day the most moving, particularly the presentation by Nyadol Nyuon, an African Australian woman, who noted that climate change is creating climate refugees around the world. She asked how Australian governments, given their poor record on welcoming refugees, will respond when millions of climate refugees seek sanctuary in Australia. Tim Costello of World Vision fame observed that although Australia has recently become the poster-child for climate change around the world, alluding to the massive bushfires that hit all six states in 2019–2020, the first victims of climate change have been the poor and South Pacific Islanders.

The most high-profile plenary of the summit focused on "Democracy Reboot: Politics Fit for the Emergency Challenge," moderated by former Australian Broadcasting Corporation journalist Kerry O'Brien. John Hewson, a former leader of the Liberal opposition, argued that Abbott did a lot of damage to climate politics and that Australia, as Ross Garnaut (2019) contends in *Superpower*, could export renewable and hydrogen energy. Peter Garrett asserted that history will judge Julia Gillard kindly, in part because the Carbon Pricing Mechanism, which the Abbott government repealed, brought emissions down. Zali Steggall argued that climate change is not a left/right issue. If that were the case, why it is that climate denialists tend to be found in greater numbers in the Coalition than in other political parties?

Of the many breakout sessions, probably the one with the most radical observations ironically was the one on "Climate Mobilisation – Politics in America." Margaret Klein Salamon (2020), a visiting US psychologist and author of *Facing the Climate Emergency*, argued that there is a need to break corporate power. She argued that Bernie Sanders is the most popular politician in the United States and that many young Americans are anti-capitalist. Christine Milne, the former deputy leader of the Greens, described the United States and Australia as plutocracies. She observed that the Greens are doing well in Germany and Ireland but expressed concern about the coalition that the Austrian Greens have formed with the right-wing Austrian People's Party, noting that there is a danger of moving into eco-fascism in Austria. Milne argued that the Austrian Greens are by default supporting the Islam-phobic policies of the People's Party which have made only vague commitments to climate change action.

The tone of the summit was generally progressive, but hardly radical for the most part. It included a breakout session on "Crisis in Country: First Nations' Responses to the Emergency," featuring Greens Victorian state parliamentarian

Lidia Thorpe, Jacqui Katona, and Neil Morris. In contrast, the breakout on "Building the Business Coalition: Getting Business on Board," featuring Ian Dunlop (a former coal industry executive), Paul Gilding, Heidi Lee, and Simon Holmes A. Court, illustrated the summit's failure to acknowledge that ultimately global capitalism, not simply the fossil fuel industry, with its focus on profit-making and economic growth, is the primary driver of anthropogenic climate change and growing concentration of wealth in the world. As Marx observed, capitalism is in a metabolic rift with capitalism.

The summit ended with the following declaration put forth by Kerryn Phelps (an independent federal MP from the Sydney area), Peter Hewson, Peter Garrett, Carmen Lawrence, Ian Dunlop, and Tim Costello, calling upon signatories to join in building leadership that embraces the climate emergency framework by:

- Promoting a non-partisan approach that embraces people of all political parties and sectors of society who are committed to science-based policies that make climate a priority of the government and the community;
- Emphasizing the value of a non-partisan government of national unity on climate;
- Holding political leaders to account if they fail to protect the Australian people;
- Prioritizing engaging with the business community to build an understanding of the real nature of climatic risks and the pace of change required to address them;
- Mobilizing civil society to take climate action;
- Supporting the formation of a specialist taskforce to set out guidelines delineating Australia's emergency transition to restore a safe climate.

Ironically, while the conference included attendees whose politics are radical in some shape or other, their voices were largely excluded in the various plenaries and breakouts. There was no input from unionists, eco-socialists, and eco-anarchists, many of whom are part of the larger climate justice movement and little discussion of just transitions. While there was some token space provided for Indigenous voices, it was largely confined to one of the many small breakout sessions in smaller rooms rather than a plenary session in the spacious main hall. While the summit aimed to engage with businesses to act on the climate emergency, representation from corporate leaders was minimal, limited to the likes of Ian Dunlop as a former fossil fuels industry leader. On March 2, 2020, Greens leader Adam Bandt introduced a bill into the House of Representatives, seconded by independent MP Zali Stegall, proposed a climate emergency declaration, requiring every government department to be guided by the declaration which mandates the creation of a "war cabinet" to address the climate crisis.

The ALP opposition and independent Andrew Wilkie (a Senator from Tasmania and a former Greens member) supported the bill. Conversely, the Liberal Minister for Energy and Emissions Reduction, Angus Taylor, pooh-poohed the motion as a grand symbolic gesture deploying emotive language. The government used its members to shut down debate. A day later Steggall printed out and tabled the largest e-petition ever received by the parliament with more than 405,000 signatures.

While in part agreeing with the spirit of various climate emergency plans, most of them are politically unrealistic and naïve. For the most part, they do not call for transcending global capitalism, merely regulating and reforming it. Given that there is a tendency for multi-national corporations in most countries to make or break governments and politicians, it seems that at least for future, few governments exist which are strong enough (albeit China may be an exception) to implement a climate emergency plan, even if they have declared them, as has the UK and over a thousand local governments. Progressive environmentalists and climate activists understandably would like to see governments, communities, and even businesses declare climate emergencies. However, as Frankel observes:

> Unfortunately, the moral and scientific case for emergency action fails to persuade most businesses and governments. The latter will never agree to drastic mitigation policies, unless there is a massive drop in the cost of renewable energy or an environmental catastrophe makes it impossible to halt mass demands for the reform of "business as usual".
>
> *(Frankel 2020: 119)*

While the threat of the Axis powers became increasingly apparent in both the West and the East during the late 1930s and early 1940s, most corporate elites and politicians, let alone ordinary people, do not presently appear to perceive the existence of a climate emergency, despite its reality, since heat waves, droughts, bushfires, hurricanes, floods, and so on, in their minds come and go. Eventually, hopefully, this perception is likely to alter as climatic crises intensify around the world.

Paul Gilding, while rejecting the growth paradigm and arguing for a drastic economic transformation, appears to leave it up to the supposedly more progressive actors in the corporate sector to nudge governments to act on climate change more action on climate change. Citing examples such as US Climate Action Partnership, which included various companies, he asked:

> [W]hat if these companies behaved as if the economy and their future prosperity and survival were at risk? What if they organized market coalition of pension funds, companies, and consumers that was so powerful, it overwhelmed the opposition from other corporations resisting change and forced government to act.
>
> *(Gilding 2011: 184)*

More recently, Gilding (2019: 23) argues that the climate emergency mobilization will be dominated by investing in renewable energy technologies and other energy-efficient technologies and electric transport vehicles, under which "some companies will not survive but will be replaced by new companies." For him, the climate emergency mobilization will come primarily from private rather than public funding, with the state notifying businesses about the gravity of the situation and merely regulating and imposing a carbon price on them.

The belief that corporate leaders, with perhaps a few notable exceptions, will do anything that interferes with their primary commitment to profit-making and economic expansion is politically naïve. Conversely, at best, they are more likely to be drawn to the mantles of "corporate social responsibility," "sustainable development," and "corporate environmentalism" as strategies for warding off public criticism. Furthermore, espousing another variant of green capitalism which entails a steady-state economy defies the reality that history has repeatedly demonstrated that by its very nature capitalism by its very nature must grow or die out. As Magdoff and Foster (2011: 56) argue, the "notion of a capitalist no-growth utopia violates the basic captive force of capitalism."

Margaret Klein Salamon appears to be more progressive, even more radical, than most of the Australians who have been promoting the climate emergency framework. She argues:

> To solve the climate and ecological emergency, we must transform our destructive economy into a *regenerative* one, and we must do it at emergency speed. We don't just need zero emissions in every sector; we need huge carbon drawdown projects that restore ecosystems and the soil. We need permaculture and food localization; we need an end to mass consumerism and endless growth; we need to give back half the Earth to nature to restore biodiversity; and we need to create a society based on protecting and healing humanity and the natural world. This means transforming not only our energy, agricultural, transportation, and industrial systems – it means transforming ourselves and how we relate to each other.
>
> *(Salamon 2020: 78)*

Salamon argues that capitalism, with its dependence on constant, its high levels of social inequality, its treatment of workers as disposable, and its heavy reliance on advertising to promote the culture of consumption, is a key part of the problem underlying climate change. She qualifies this observation by noting that the Soviet Union was approximately as damaging to ecosystems as the United States and was the second-largest emitter of greenhouse gases during the 1960s, 1970s, and 1980s, and that China's largely state-driven economy has become the world's largest emitter of greenhouse gases. However, from an eco-socialist perspective, it is important to note neither the Soviet Union nor China were authentic socialist societies, a reality that has thus far never been achieved in any country for complex internal and external contradictions and historical and social structural reasons (Baer 2018). Indeed, while China may have constituted a post-revolutionary or socialist-oriented society during its Maoist period, some have argued that it has evolved into a state capitalist society.

Even the eminent Slovenian philosopher has adopted the climate emergency discourse, although it is difficult to say from where he picked up:

> [G]lobal warning effectively *is* a (not only) emergency. Proclaimed or not, we *are* in a state of emergency. In view of the latest news about global warming, which makes clear that it progresses much faster than even the pessimists expected, some commentators quite appropriately draw a parallel with World War II – nothing less than a similar global mobilization is needed.
>
> *(Zizek 2020: 84)*

At one level, governments, whether local, state/provincial, or national, or the United Nations' declaring climate emergencies is well and fine, but the larger issue is the nature of the actions that will be taken to address them. Will they be actions that seek to work within the parameters of global capitalism or will they be actions that seek to transcend it? The UN Framework Convention on Climate Change has made numerous commitments to reduce greenhouse gas emissions, the 2015 Paris commitment being the most notable one, but emissions continue to rise, now standing 415 CO_2 ppm. There is a need for the climate emergency movement to become part and part of a larger climate justice movement, not simply a climate movement that emphasizes techno-fixes, and that says, "system change, not climate change." For a growing number in the climate justice movement, the system change needed entails a socio-ecological revolution shifting the capitalist world system to an eco-socialist world system.

A life history of Rod Quantock: a renowned Australian left-wing comedian and climate activist

This section presents the life history of Rod Quantock, a renowned left-wing Australian comedian with over 50 years of experience in stand-up comedy, cabaret, theater, television, and radio. He is considered the grandfather of stand-up comedy in Australia and is the only comedian to have appeared in all 33 Melbourne International Comedy Festivals. His work has always been critical and satirical. Shows with titles like *The Axis of Stupidity*, *Baghdad Nights*, *Scum Nation*, and *Boredom Protection Policy* have examined the Iraq War, neoliberal politics, the media, money, refugee policy, and more. When Rod is not working as a freelance writer, performer, and speaker for the public, nonprofit and private sectors, he is a prominent environmental, social, and political activist. In 2015 he was awarded an Order of Australia Medal for his contribution to the performing arts, the environment, and sustainability. Rod has held positions as an honorary associate in the Melbourne Sustainable Society Institute and as a research fellow and honorary research fellow in the School of Culture and Communication at the University of Melbourne. He was the first comedian in the world to tackle climate change and the limits to growth, and over the course of the past 15 years, he has written and performed exclusively about them. With titles like *Bugger the Polar Bears*, *This Is Serious*, *The*

People We Should Eat First, Pardon My Carbon, Peak-a-Boo and Catch CO₂, A Brief History of the End of the World, and most recently, *The Last Tim Tam*, he has dissected politics, economics, globalization, militarism, and what he terms the "cargo cult of corporate capitalism."

Rod has read the literature, watched the documentaries, interviewed scientists and experts, and has an encyclopedic knowledge of climate science. He also understands that climate change is not the only issue. Rod also talks about peak water, peak oil, peak food, peak money, peak sand, peak soil, peak everything, and relevantly, pandemics.

In 2008 Rod published a booklet titled, *Rod Quantock's Guide to the Wonderful World of Climate Change* (Quantock 2008). Various people have referred to him in conversations as a "national treasure," although undoubtedly many people of a conservative or even a middle-of-the-road orientation might dispute this designation. On the matter of climate change, Rod, based upon his observations and experiences, has adopted a dystopian or catastrophist stance, a position that has prompted me as a proponent of democratic eco-socialism as a real utopian response to the socio-ecological crisis to enter into a dialogic conversation with him.

Methodology

The life history method has been an integral part of anthropology since the 1920s, particularly initially focusing on Native Americans (Langness and Frank 1981; Runyan 1982). My discussion of Rod's childhood years, his university years and entrée into the world of comedy, and his becoming a left-wing comedian and political activist conform more closely to the traditional life history approach. Conversely, the section on his worldviews, particularly with respect to capitalism and the culture of consumption, politicians and political parties, climate change, and possibilities for changing the system, conform more to the life story approach.

Life history traditionally has been a methodology utilized particularly by anthropologists to capture through one individual the essence of a culture or lifestyle, such as the case of Paul Radin's (1920) classic book titled *Crashing Thunder*, an account of a Winnebago Indian.

Over the years, in my role as a scholar-activist, I also have attended numerous events featuring Rod, including a climate rally to close the Hazelwood coal-burning power plant in the La Trobe Valley in 2010, several *Green Left Weekly* comedy debates that he moderated, four performances of his *Last Tim Tam*, and a Melbourne Comedy Festival performance marking his 50th anniversary as a comedian. Given that for a period of time our offices were situated in the same building at the University of Melbourne, we had the opportunity to chat over lunch or coffee. On one of these occasions, I asked Rod if anyone has ever written a biographical sketch of his career as a stand-up comedian. He noted that someone had written a Wikipedia article about him, but otherwise, no one had written something along these lines. When I told Rod that he would make an interesting subject for what anthropologists term life histories, he expressed interest in my offer to do one on him. Following the tedious process of obtaining an ethics clearance for the project, our interviews commenced.

Between April 7, 2018, and January 24, 2019, I conducted six interviews with Rod for purposes of a life history/story of an icon in both comedy and Australian social activism, one who provides a glimpse into the social universe of Australian left-wing political activism. For the first five interviews, I posed a long list of semi-structured interview questions to Rod about a wide assortment of topics, such as his upbringing, studies as an architecture student, social activism, involvement in political protests during the late 1960s, his career as a professional comedian (both on the stage and in TV advertisements), a stint as a weekend journalist for *The Sunday Age*, and his evolution into a climate activist. In late 2018, Rod conducted a 30-minute rendition of his *Last Tim Tam* performance at the double book launch of the second edition of *The Anthropology of Climate Change* (Baer and Singer 2018) and *Democratic Eco-Socialism as a Real Utopia* (Baer 2018) in the School of Social Sciences at the University of Melbourne. Our sixth interview took more the form of a dialogic conversation around Rod's climate change dystopian views and my own vision of democratic eco-socialism as a real utopian response to the socio-ecological crisis. For this purpose, he read twice my Next System Project essay on democratic eco-socialism, which essentially is an abridgment of my book (Baer 2016).

Childhood years

Rod was born on August 24, 1948, lived in Springvale, at the time a semi-rural fringe suburb in south-eastern Melbourne, for the first one and a half years of his life, but grew up primarily in Coburg, a northern inner-city working-class suburb of Melbourne. Before World War II, his father, Ron, a fourth-generation Australian, worked primarily as a metal polisher adding the shine to car parts, silverware, Football Grand Final Cups, and the Australian television equivalent of an Oscar, Logies statues. After the War, he became a tram conductor but returned to metal polishing after an accident. Ron's father had abandoned the family when he was young and he had to forgo his education to help support the family. (He only completed second form, something he regretted the rest of his life but it made him ambitious for his children's education.) He was an enthusiastic reader who exclusively read cowboy and detective novels, but he passed on his passion for books and education to Rod.

Rod's mother, Lorna, was a war widow who married Ron, a divorcee, at the end of World War II. She did subcontract work at home folding and packaging shirts for a clothing manufacturer, a task which Rod and his siblings assisted her with at night. Rod surmises that his father voted for the Australian Labor Party (ALP), but he did not discuss politics at home, at least until the Vietnam War, to which, because of his generation's experience of war, he was vehemently opposed. Rod said that his mother was not politically well-informed and may have voted Liberal to annoy his father. His father took him on weekly trips to the local library where Rod borrowed encyclopedias and history books. At the weekend Rod took himself to the Melbourne Museum or the Zoo and the local picture theatre. Television came into Rod's home in 1958, and he vividly remembers watching pioneering oceanographer Jacques Cousteau's series about life in the oceans and considered a career as a marine biologist. "Before I was 10 I had seen B&W newsreel footage of *Hiroshima*, *The Burma Railroad*, *Kakoda*, *The Holocaust*, *The Blitz* and all the horrors and heroics of war." The 1962 Cuban Missile Crisis was a seminal moment for Rod. The whole world was trembling. He says his school woodwork project that year was to design and construct a plywood fallout shelter. Rod observed,

> When I was 16 I saw Peter Watkin's Academy Award-winning docudrama, *The War Game*. Commissioned by the BBC to commemorate the 20th anniversary of Hiroshima, it was a brilliant, brutally realistic and terrifying depiction of a nuclear attack on Great Britain. It sat in my mind next to Stanley Kubrick's "Doctor Strangelove."

He added,

> As a set text in my final year of high school, I read *Brighter than a Thousand Suns: A Personal History of the Atomic Scientists* by Robert Jungk. The book was the first account of the Manhattan Project and wartime attempts to develop an atomic bomb. It made me realize that scientific discoveries can and will be weaponized.

It reminded Rod of God's warning not to eat the fruit of the Tree of Knowledge. He notes that many technologies – GPS, drones, materials, machines – have been born in military research projects. Rod observes,

> All of this gave me a respect for the staggering achievements of Science, a respect amplified by my inability to master any aspect of it. But I also recognized that the worst in us is perfectly capable of destroying the world with the power that Science reveals.

Rod's parents were not church-goers, but they made their children attend weekly Sunday school, each taking a penny for the plate. Somehow Rod's penny was miraculously transformed into a halfpenny for God and a halfpenny bag of mixed lollies for Rod. He observes, "My religious upbringing was very ecumenical: I played football for the Catholics, cricket for the Methodists, basketball for the Salvation Army and went on picnics with Baptists." In 1954, Rod made his stage debut as the Guiding Star in the West Coburg Methodist Church's Nativity Play. "I just stood up the back and twinkled over Bethlehem," Rod recalled, adding: "When I waved with my star to my parents I got a laugh. A comedian was born." Rod says that for many years he had his own idea of a non-institutional god but now sees existence as an interplay of energy and interdependence. That is not to say that religions have no value, he says. For him, they are ways of trying understanding Life and Living but they too easily become ways of controlling and dividing. He talks of a Hindu idea that the cosmos is a net with no beginning or end. Each node of the net holds a jewel, and that jewel reflects every other jewel. Everything is related to everything else. That, he says, is his idea of socialism. He also maintains Jesus was a socialist.

Rod's father regularly took his family on day trips to the Mountain Ash forests east of Melbourne. He observes,

> The Mountain Ash forests of Victoria are lush and majestic. They were home to the tallest tree in the world. It was cut down in the 19th Century and all that remains is a sign in the middle of dairy pastures to mark where it was.

Rod laments the loss of the natural world, observing "I've witnessed the relentless expansion of suburbia, devouring everything before it." He argues: "We are the only creatures that have ever existed that can interrogate nature, understand nature, see the beauty of nature, enjoy nature and, in a so exceptionalist way, destroy it. Nature is something we are part and without it we are nothing."

Rod thinks people have always feared nature, its power and cupidity. It can be life-threatening. It can be cruel in the way that it gives and takes. Through ritual and technology, people have tried to appease and tame it. He argues, "It's chaotic and unruly. I had a neighbour who used to vacuum her lawn clippings with a domestic vacuum cleaner. When the autumn leaves fell, she vacuumed them." Rod says,

> [The] "invention" of agriculture 12,000 years ago, was the point where everything changed. Up until then, we were part of nature, of it and in it. We were hunter-gatherers; we didn't have deep impacts on the environment. But we learned to domesticate animals and plants and store grain. Agglomerations of people settled around farms and wanderers became permanent residents, breaking the soil and fencing nature out. Sedentary people began to have personal possessions, property. What once was shared was now "mine". Over time villages became towns and towns became cities. Hierarchical political and social structures, divisions of labour, privilege and exploitation evolved. Walls went up to keep others out. Bloodied empires rose and fell. And we stopped fearing nature and began to fear ourselves.

Rod attended Moreland High School, which was one of the many schools throughout Victoria sold off in the 1990s by the neoliberal Kennett government. Victoria was one of the early adopters of Thatcherite privatization and a feeding frenzy ensued: "Utilities, public transport, public lands, government services were put on the auction block." Rod had some excellent and pioneering high school teachers, including a history teacher who had graduated from the University of Oxford, an English teacher who went on to become a senior administrator in the state education system, a math teacher from the United States, an English Literature teacher who became head of the Canadian state education system, and an economics teacher and future academic who gave him second-class honors for his efforts. All were left-leaning. Rod also played football and golf during his high school years, delivered newspapers, and admits that he had a "pretty easy-going life." However, the seeds of Rod's political awakening were planted while still in school.

Rod was also exposed to the larger world through the news clips at the movies he viewed at the local picture theatre and Life Magazine to which he subscribed. He particularly became fascinated by Watergate and the US civil rights movement, but he admits that, while he knew of the treatment of the first Australians, he did not become fully aware of the indignities that White Australians meted out to them until the emergence of the Australian Indigenous civil rights movement in the 1960s. He delivered morning newspapers and was one of the first people to learn about the assassination of John F Kennedy when he delivered the headlines to a sleeping world in 1963.

University years and entrée into the world of comedy

Upon completing high school at the end of 1967, Rod spent the summer manufacturing and laying concrete blocks before enrolling in a Bachelor of Architecture course at the University of Melbourne. He became fascinated with Melbourne's iconic architecture. Rod feels that the powers to be in Melbourne have not appreciated the architectural history and heritage of what started out as "Marvelous Melbourne" and have been "very cruel to the city," transforming much of it into a "developers' paradise."

A first-year project examining aspects of urban development gave him the idea of showing fellow students, many from privileged leafy-suburbs backgrounds, how the other half lived. He hired a bus and took them on a tour of an impoverished urban fringe public housing development where the concrete block factory he had worked in was located. It was on a barren dirt road lined on one side by asbestos-sheet houses and on the other by factories including a dog food factory that blanketed the area in a Dickensian stench.

To ensure the students got a thoroughly immersive experience, Rod and his mates sent the bus away so that the students would find themselves stranded, thus requiring them to use the inadequate public transport of those neglected fringe suburbs to get back to the University. However, privilege prevailed once again and the well-heeled students found a phone booth and called for a taxi to bail them out.

Like many now older Australian radicals, Rod was shaped by the social turbulence of the 1960s. As Gerster and Bassett observe,

> The sixties was a time of "dirty words", words like "authority", "capitalism", "Establishment", "status quo", "conservative", "sexism", "racism", "pollution", and the more conventional expletives uttered in defiance of censorship. But the dirtiest word of all was Vietnam.
>
> *(Gerster and Basnett 1991: 42)*

As in the United States, Australian "anti-war protests were centred on the campuses" (Gerster and Basnett 1991: 44). In 1967, the very same year that Rod began his university studies, Australian students began to engage in militant demonstrations against the Vietnam War, for which young Australian men had been conscripted. Hastings reports:

> They marched on the streets and, in 1968, threw stones at the US consulate in Melbourne. On some campuses, a particularly politicised minority openly collected money for the Vietnamese National Liberation Front and set up campus sanctuaries for draft resisters.
>
> *(Hastings 2005: 101)*

While as an elite sandstone university, the University of Melbourne did not feature as much radical activity as did Monash University and La Trobe University, it had its share of protest activities, including against the Vietnam War. Rod was conscripted in his first year, although his call-up was deferred until he graduated, which he never did. A change of government in 1972 ended conscription. Rod feels that his time at Melbourne University politicized him. A raid by the Australian Security and Intelligence Organisation (ASIO) that smashed into the University Student Union Offices looking for draft dodgers and antiwar activities alerted him to the desperate ruthlessness of political power.

Rod thinks the Sixties "put the wind up the ruling classes," who, he says have some little part of their brains haunted by Madame la Guillotine. The decades since have seen a well-funded propaganda assault on the ideas and institutions of the progressive left to justify small government, deregulation, attacks on universities, and the rights of corporations to have the liberties of persons. Rod works from a fundamental premise that, invariably, the Left is right and Right is wrong, He feels that has served him well in a complex world.

Rod loved studying architecture because it provided him with insights into politics, history, and religion. In his view, buildings are the public face of civilization. Architecture leaves permanent reminders of the rise and fall of empires, the place and nature of religion in society, and the relationships of the ruled to the rulers. Feudal castles tell of an economic system that Rod sees echoed in the working of contemporary capitalism. The slums of Glasgow are reminders of the brutality and indifference to the suffering of the masses in the pursuit of wealth. The sprawl of suburbia tells the story of oil and the motor car. While Rod enjoyed designing and constructing models of buildings, he struggled with physics, chemistry, and math. Later in life, he questioned his failure and began reading histories of the sciences and mathematics. It didn't make him any more competent, but it resulted in 20 years of talking in schools on the history of mathematics. It also gave him the grounding to engage with climate change. Furthermore, Rod wryly maintains that he discovered that he would probably spend his life as an architect designing house extensions rather than palaces and museums, prompting him to drop out of the six-year architecture program at the end of his third attempt at third year. In the end, architecture's loss turned out to be comedy's gain because by this time Rod had discovered his forte in student theater and student politics. Along with other students at Melbourne University, he marched in the demonstrations against the Vietnam War and protested against the demolition of historic buildings.

The making of a left-wing comedian and political activist

After abandoning his architecture studies, Rod worked for three years as a draftsman for a firm that, ironically, made concrete products. But his first love was comedy, and he became involved in writing, performing, and directing independent off-campus shows with a group of other Archi Revue alumni. In 1974 John Pinder, a New Zealand-born, lifelong pioneering producer, opened the Flying Trapeze Café and Theatre Restaurant in inner-city Fitzroy, a venue licensed for 40 people but regularly accommodating 80. As a comedian, however, the big money came not for his comedy but for his 20-plus years advertising contract for a bedding company, Captain Snooze. Rod says that while the contract gave him financial security, something that performing artists rarely have, Captain Snooze gave him the "wrong kind of publicity," despite the fame that the role brought him. Rod's last ad for Captain Snooze was in 1996. The bedding company asserted that it let him go because of a "changing demographics." However, he thinks that his very public political activism was the underlying reason for his dismissal, noting, "You have to be politically neutral in advertising. Very few celebrities take sides."

In the early 1990s Rod became heavily involved in a campaign to stop the destruction of one of Melbourne's iconic city-fringe heritage public open spaces, Albert Park. The Coalition Kennett government began the destruction of large areas of the park for the carpetbaggers of the Formula 1 Motor Racing Grand Prix. Rod was an important part of the campaign protesting, primarily as a "foot soldier," but using his public profile to highlight the issues. During these protests, he moderated some campaign fundraisers, and, not for the first or last time, was manhandled by the police. It was his very public involvement in this campaign and others at the time that he credits with the loss of his bedding commercials. Rod attended the protests two or three times a week for a long time, but despite the efforts of the best-run community protest campaign to that date, the park was mutilated, and the Grand Prix went ahead and continues to this day. The experience taught him a bitter lesson: "You lose more campaigns than you win." Rod notes that within six months of Jeff Kennett's election as Victorian premier, over 350 protest groups, many of which he gave his time to and publicly supported, had formed in Victoria, most mounting single-issue campaigns such as "save our schools," "save our park," or "save our post-natal facility." Rod quickly realized that the formation of so many protest groups inadvertently functions essentially as a "fantastic way of splintering resistance" to the system: what could be a conflagration to destroy the system becomes a series of spot fires. Conversely, though, campaigning creates community.

Despite Rod's skepticism about the effectiveness of many protest efforts, he does admit that there have been some success stories in Melbourne, such as the effort to stave off developers in Clifton Hill, his home suburb, and the stopping of the privately operated East-West Link toll road, crediting particularly socialist groups that sponsored a campaign caravan that traveled around Victoria as part of their anti-East-West Link actions.

Rod has also been a strong supporter of the union movement and recognizes their great contributions not only to the interests of their members but to the broader community. He says they added their weight to the preservation of heritage buildings and precincts, have supported the struggles of workers around the world, and have opposed wars and corporatization. But Rod says constant anti-union propaganda legislation has gutted much of their efficacy. He observes the best thing about all those protests has been the remarkable people he has met, the things they have taught, and the friendships he has made.

The world as seen by Rod Quantock

While Rod has made his living as a comedian, during the 1990s he turned some of his energies to satirical journalism, writing a column in *The Sunday Age*. (It ran for five years until it was abruptly terminated because, Rod says, of his many critiques of Premier Jeff Kennett.) Some of his articles appear in a compiled collection titled, *Double Dissolution: Musings from a Rebel Comic* (Quantock 1999). Themes covered in his articles include employment and unemployment, the wool crisis, Bob Hawke, Jeff Kennett, Kerry Packer, Australian banks, the ABC, marriage and divorce, the government budget, Melbourne heat, the Vietnam War and the War on Iraq, Australia's support for Indonesia in the conflict with East Timor, the Olympics, the AFL Grand Final, Father's Day, Rupert Murdoch, Manning Clark, Princess Diana, artists and art, opera, the Victorian state election, Melbourne architecture, Melbourne University, the Tullamarine and Citylink Tollways, Coalition Party Federal Treasurer Peter Costello's ideas on tax reform, and the General Sales Tax (GST). One example of the type of satirical commentary in his numerous articles is captured in his comments on his alma mater:

> In the middle of 1999 Melbourne University knocked down two adjoining Victorian two-storey terraces on Carlton Square, opposite the University. They didn't knock them down lovingly, with reverence and care, salvaging the lace and timber trims. They crushed them flat in 20 minutes lest the people of Melbourne should gather to save them. Melbourne University did this just weeks after McDonald's behaved in a similar rapacious, corporatist way when they destroyed a 100-year-old house in Coburg [Another example of Rod campaigning and loosing]. When a prestigious university, a repository of civilized learning, can do that I wonder where education is going.

Rod feels that his three years – five if you count repeating the third year twice – as an architecture student at the University of Melbourne were the best years of his life and was thrilled to become involved with academia again in 2016, where he worked researching ways of communicating climate change and limits. However, he feels that there are people at the university who do not like what he has to say particularly about climate change and its drivers. In contrast to his student days, he maintains that student demand drives the university today and critical thinking has diminished in the university. Rod argues:

> Universities are big business. They sell courses to people that are increasingly vocational. They are part of the system now, but they probably have been for a long time. When universities put slogans such as "Dream Large" or "Believe" on billboards, they inhabit the world of Nike. And, like all participants in markets, they tread warily when criticising the hands that feed them.

Rod cites the case of urban planner and internationally renowned public transport expert, Professor Paul Mees, who lost his position at Melbourne University. Various newspapers, including the *Sydney Morning Herald* and *The Age*, covered the story and said, "[H]e was told his pay would be slashed and his position downgraded after he made a strongly worded attack on the Government over transport privatization. He has since resigned." The University of Melbourne lowered Mees' rank from senior lecturer to lecturer, but he managed to recoup himself by obtaining a position as a senior lecturer at the Royal Melbourne Institute of Technology. Rod laments the low level of student activism on campus, but he admits that this may be related to the need for many students to work one, two, or even three jobs to pay for their education. Students do courses online and staff lecture to empty rooms. Students rarely attend a campus. The foment of post-war student politics has, as a consequence of attacks on higher education, been diluted.

Capitalism and the culture of consumption

Of capitalism, Rod notes, that while he is not an anarchist per se, he is unequivocally anti-capitalist, asserting that "if someone said that we are overthrowing capitalism tomorrow at 7:30 am, I would be there." He adds:

> My parents' generation had political aspirations fired by their experiences of a Great Depression sandwiched between two World Wars. That generation demanded the eradication of poverty, the implementation of universal welfare, state ownership of essential industries. Unfortunately, the past 50 years have seen Australians trade those aspirations in for bread and circuses.

Rod contends that Australia has relinquished its sovereignty to corporations, entities that are devoid of any attachment to nations and their peoples, turning societies into markets and the natural world exploitable resources. He blames the overall high standard of living in Australian society for widespread political apathy. Rod maintains, "It is hard to move people who are happy with football, BBQs, and a glass of wine."

Rod asserts:

> People live in headset bubbles, listening to music wherever they go. And nothing seems to get in except what they want to get in and that's the worst of it now. The Internet's made the world a place where you only see and hear the things you agree with. Nobody looks up contrary views because they're a challenge to their ideas and the way they've constructed their own personal worlds. Societies are atomized and incoherent. There are conspiracy theories and conspiracies, propaganda and ubiquitous surveillance, data mining and targeting. Self-interest trumps truth.

He adds that the only way people will respond to climate change is when it floods their home, burns it down, or blows it away.

Politicians and political parties

Rod has no interest in politics anymore, arguing, "Australia – and the USA and the UK – are just one, giant enormous arse, and it's called Rupert Murdoch, and everybody is up [to] it. Murdoch's empire controls the messaging on behalf of what Rod calls, feudal capitalism. Trump and Brexit are his children." Rod notes that things are marginally better here because Australia has compulsory voting and a higher, but diminishing, economic equality. He maintains the coronavirus (COVID-19) has not only exposed the weakness of nation-states in a global, inequitable world but laid bare the self-interested and psychopathic delusions of the powerful and that the Murdoch press has been at the forefront of the "end the lock down" campaign. Rod feels that it has downplayed the lethality of the virus to get capitalism working for capitalists again. Their arguments advocate falsely that younger, working people don't die from the virus so get them back to work, while the vulnerable old and infirm should be willing to take a viral bullet for their children's economic futures. He notes that the ruling class has always been willing to immiserate and destroy the lives of the masses to build their wealth and power; from the slaves who built the pyramids to the child labor of the Dark Satanic Mills of Asian sweatshops, it has always been thus. Rod maintains: "Partisan politics in corrupted 'democracies' cannot address the impending collapse. The time to do that was 40 years ago."

Rod said, as part of his research, he spoke to the former head of the UK's Hadley Centre, one of the top three climate research bodies in the world. Rod asked him what could be done to keep warming below 1°C (the first of the ever-escalating "safe" benchmark temperature), and he said, "Nothing." Rod adds that the much-celebrated 2015 Paris Accord commitment to hold global temperature rise to 1.5°C is like him committing to hold the lead in the 1980 Tour de France: "Nice idea, too late." Rod observes, "We're well past 1.5 degrees that everybody talks about. There are days, in fact, weeks now where the global average temperature is two degrees above the average."

Climate change

Rod has been concerned about the environment for as long as he has been concerned about social justice. Initially, his environmental concerns focused on specific issues: a forest, a tree, a beachfront, a park. His view is now systemic. In 2007 Rod received a two-year grant from the Theatre Board of the Australia Council to write a history of the world from 1948, the year of his birth, to 2008, his 60th birthday. During his research, when he got to the oil crisis of the early 1970s and the concept of climate change, he started to think about the limits to growth. Consequently, Rod published a booklet titled *Rod Quantock's Guide to the Wonderful World of Climate Change*. In that first iteration his prediction for the year of the collapse was 2050. In 2016, he revised the date 2030, which he sees as a "realistic line in the sand, for when things with the climate will really go wrong." The guide is a satirical piece of work that in words, pictures, and graphs illustrates the gravity of climate change for Australia and humanity. Mock front pages of future editions of the Melbourne Age newspaper run headlines like: August 3, 2030 "Lilydale & Ringwood devastated: the worst bushfires in Victoria's history ring Mooroolbark as temperatures hits 58°C" and for February 13, 2050, "Civilisation collapses: PM promises to rebuild" and on February 10, 2070 the headline reads, "Mercury climbs above seventy' and 'last people leave city."

In Rod's apocalyptic view of the climate crisis driven by fossil fuels and a culture of insatiable consumption, his guide reports:

> The world that sustains us is shrinking as we grow in number and wants. The water, the food, the very ground we stand upon is vanishing. We are consuming ourselves and all living things out of house and planet.

> Sometime, very soon, an invisible molecule of carbon dioxide or methane (Watch it cows!) will be the invisible molecule of carbon dioxide or methane (You too frozen tundra) that tips the planet's climate over the edge.

> Life itself may all but disappear – everything from worms to whales gone in perhaps three generations and with them goes us and our world. No more AFL or ice-cream, no more pizzas or plasmas, no more everything . . .

> It's time to COME CLEAN. We have two years to STOP THE MADNESS.

Given his long-time interest in mathematics and science, Rod began to weave the findings of climate science into his performances, particularly "The Last Tim Tam," which satirically argues that catastrophic climate change will make the resources that go into the production of Australia's iconic chocolate-covered biscuit unattainable. He notes that the emphasis on conventional solutions, such as renewable energy and market mechanisms, promoted him to produce "The Last Tim Tam," a performance that he has presented numerous times in a wide array of venues, each time improvising a bit to account for the most recent climate scientific observations.

Rod thought if he told enough people about the gravity of anthropogenic climate change, Australians would see the need for climate action, particularly in the form of mitigating climate change, but to a large degree, he does not think his audiences have taken his concern seriously. In fact, Rod claims that he has learned not to discuss climate change in certain circles. He maintains that the global ecosystem has become utterly chaotic. He says that oceans and forests will soon be emitting carbon dioxide rather than sequestering it. Rod asserts that many people don't want to hear that capitalism is the ultimate driver of climate change because many of them are invested in the system in one way or other. One of the great stumbling blocks of capitalism is a public company's legal duty to maximize shareholders' profits. But he notes there are now legal challenges for companies to accommodate the cost of climate change to their bottom lines.

Rod argues that capitalism is a paradigm of the Third World supporting the rich world, noting that "we have ten-year-old children who wear clothes made by ten-year-old factory workers in the Third World." Rod observes that capitalism is a system that has "claws into everything." He feels that the wide array of environmental groups addressing climate change are wasting their energy focusing on primarily technological solutions: "It is technology, powered by 'limitless' energy that got us here and by its very nature is incapable of turning back the clock."

Changing the system

Rod feels passionate about many causes – the sanctity of natural world, economic equity, human rights, and individual dignity – and always operated on the belief, that while it takes time, things do improve. He feels that climate change, the desecration of the natural world, and the triumph of unfettered free-market capitalism and its capture of the political process has the world one unavoidable step away from The Sixth Great Extinction.

Rod says,

> I began to think the worst 15 years ago. But being a comedian not a scientist, I thought "What would I know?" So I did three years of research, checking with scientists and experts on the way and turned out I got it right. I spent the next dozen years looking for solutions but, guess what, there aren't any. It is a wicked problem and wicked problems have no solutions. Politics of any persuasion can't solve it because it requires everyone, particularly the affluent of the world, consuming substantially less and that is not going to happen in a planned way because no politician is going to campaign on "I will give you less". The success of democratic systems is always founded on, "I will give you more". Reducing emissions won't work because there are too many already. From a pre-Industrial baseline of 280 ppm of CO_2, today it is 415 ppm. Restoring the health, abundance and ecosystem services of the natural world is impossible. The end is finally nigh!

When the inevitability of catastrophe first dawned on him, he said every time he saw a child, he'd literally cry because he knew that child's future would be nothing but a Hobbesian struggle for survival and an increasingly deadly world.

Rod says that, in a way, he's sorry he ever pursued his curiosities. He wonders if it would be better that he didn't know what he knows and he struggles with the ethics of "spreading the news" through his work. "When there are no solutions is it right to drag people into my world of existential dread?" However, he becomes dejected when he meets highly educated people, such as mainstream economists, asserting that climate change can be contained with market mechanisms. Rod admits that perhaps we need a "severe form of socialism" or even a "pleasant form of socialism" that transcends capitalism to effectively mitigate greenhouse gas emissions. But he said: "Given the time frame – as little as ten years – and absent Madame la Guillotine, it isn't going to happen, so he sees no reasons to pursue any broad systemic solution."

While Rod over the course of his career as a comedian has attained a certain level of fame, some might say notoriety, over the past couple of decades or so, he notes that he does not receive as many invitations to perform as he did in the past and that many younger audiences do not know about him. Over the years Rod has received numerous awards both for his performances as a left-wing comedian and as a social activist. One of the awards that he values the most is the Archbishop Romero Social Justice Award presented by a small Adelaide-based group headed by Sister Janet Mead, who recorded a version of the Our Father in the 1960s that made a worldwide impact and provided her with funds to support community gardens and homeless people. Rod also received the Sidney Meyers Performing Arts Award in 2004. He feels humbled that he received the Order of Australia Medal in 2015.

Rod Quantock live

Rod says that commercial TV programs function to fill in the gaps between advertisements and do not want to offend the sponsors. He maintains that while "everyone makes jokes about the Prime Minister," few comedians genuinely critique the system. On April 7, 2018, I attended one of a series of "Rod Quantock: Happy Birthday to Me" performances at the Melbourne Arts Centre to celebrate his 50th year as a comedian. He crawled on all fours onto the stage and said, "This is a show about how we are all going to die." As a footnote, Rod added if there are solutions, they are local self-sufficiency with no dependencies on the outside world. He has thought about buying into one of those super deluxe bunkers that multi-millionaires have. But aside from not being able to afford one, he said,

> I wouldn't like to spend the rest of my life in a hole with a lot of rich pricks who we're happy to profit from the world burning and who hope that by the time the doors open, everybody else will be dead. They will be, but so will

be the capacity of a dead planet to support them. There will certainly be no toilet paper, or stoke market, or pool boys . . . or Tim Tams.

Reflections on comedy and ethnography

In addition to having come of age more or less during the same era, namely the Sixties, he in Australia and I in the United States, Rod and I started to become radicalized during this decade and the subsequent decade. Whereas he shifted from studies in architecture to a career as a comedian, I shifted from work as an engineer in the aircraft industry to a career as an anthropologist. For both of us, our respective second careers so to speak were part and parcel of our attempt to comprehend our respective societies and the world system.

What Rod as a comedian and I as an anthropologist share is that we are both climate change communicators, albeit he through the lens of a left-wing comedian and I through the lens of a critical anthropologist. Drawing upon his examinations of the climate science literature and conversations with climate scientists, Rod imparts his knowledge about anthropogenic change both through stand-up comedy and his satirical guide to climate change (Quantock 2008).

In my extensive work on my own and with others, I have attempted to communicate about the gravity of climate change and ways to address it through radical social action (Baer 2019). Along with Merrill Singer, with whom I have collaborated since 2009 in an effort to create a critical anthropology of climate change, I maintain that

> anthropologists and other social scientists can play a small but critical role in providing their analytical skills and insights to a much larger struggle to create a world in which we learned to live in harmony with one another and the planet, and in the process create a safe climate.
>
> *(Baer and Singer 2018: 219)*

Conclusion

As a climate justice activist, while I have often, both in print and in dialogue, argued for the need for climate activists to incorporate a stronger sense of social justice and critique of capitalism, I have opted to struggle to function within the bowels of the larger Australian climate movement. As Canadian eco-socialist Ian Angus states:

> There are far more people from non-socialist backgrounds than there are of us. We won't always agree with specific actions or slogans or demands, but that's just how it's going to be. Standing on the sidelines criticizing will get us precisely nowhere: socialists must be in the movement, building it to the best of their ability. . . . We must give priority to fighting fossil fuels because that's where such a movement can actually have a substantial impact, even

if we can't change the entire system yet. If we can't shut down a pipeline or prevent fracking someplace or get a university to divest itself of investments in the oil industry, how can we imagine that we're actually going to over-throw capitalism? A socialist movement that doesn't take defending human survival as a central goal isn't worthy of the name.

(Angus 2017: 163)

Nevertheless, seeking to work as an avowed eco-socialist within the Australian climate movement can be a daunting task. I often hear climate activists in Australia say that we do not have enough time to transcend global capitalism to be able to create a safe climate for humanity. Thus, they argue that climate activists need to collaborate with more supposedly progressive corporate leaders and politicians in tackling the climate crisis within the parameters of the existing global political economy. In my view, combating climate change and global capitalism need to be merged. While the more enlightened corporate elites and their political allies, such as Al Gore, may permit some measures that contribute to climate change mitigation, they will certainly not consciously permit the eventual demise of global capitalism and the emergence of an alternative to it. Despite certain weaknesses in the Australian climate movement, I remain committed to struggling within it, both in the academy and outside of it. Along with others, particularly those who take a strong social justice stance and critique global capitalism, I hope that in time, it will transform from a climate action movement to a climate justice movement and form linkages with similar efforts elsewhere, in both the Global North and the Global South.

6

TOWARD ECO-SOCIALISM IN AUSTRALIA

I start out this chapter with a dialogic conversation that I had with Rod Quantock during our sixth interview concerning the gravity of the climate crisis. He made a case for climate catastrophism, arguing that it is too late to avoid a global climate catastrophe. Conversely, while admitting the climate crisis appears very grim with no serious solution in the works at the moment or in the immediate future, I made a case for eco-socialism as offering the ultimate climate change mitigation strategy. In the following section, I consider ecological modernization as a hegemonic mainstream perspective for mitigating climate change followed by a discussion of various radical perspectives, namely eco-anarchism, eco-feminism, the de-growth paradigm, and an Indigenous perspective as they have manifested themselves in Australia as post-capitalist strategies for addressing both the ecological and climatic crises. I follow with a discussion of the role of anti-systemic movements and pre-figurative experiments in creating a socio-ecological revolution. Finally, I discuss 11 system-challenging steps for implementing a socio-ecological revolution in Australia, one that is conversant with eco-socialist strategies in both the Global North and the Global South.

A dialogic conversation with Rod Quantock: climate catastrophism versus eco-socialism

I found myself troubled as my interviews with Rod proceeded because he had come to embrace a catastrophist stance regarding the implications of climate change for Australia and humanity, that everything is doom and gloom. Indeed, Mike Hulme (2013: 383), a renowned climate scientist and climate change communicator, goes so far as to argue that the "discourse of catastrophe is in danger of tipping society into a negative, depressive and reactionary trajectory." While Rod admits to having felt depressed about the gravity and implications of climate

DOI: 10.4324/9781003202530-7

change for humanity, I would by no means characterize Rod's catastrophist stance on climate change as reactionary in that his shows as I have personally found do not generally cater to an audience consisting of a large number of climate denialists or skeptics, but more so among people on the left, including the far left. Having seen his climate change performances on several occasions, I suspect that he does impart a sense of fear among some people about the gravity of climate change and the capitalist-consumerism that drives it. Whereas Hulme (2013: 382) contends that the "language of fear and terror operates as an ever-weakening vehicle for effective communication or inducement for behavioural change," I find myself in agreement with the following statement made by psychologist Margaret Klein Salamon:

> There are some obvious problems with the implicit decision made by scientists, the climate movement, and the media to avoid scaring people. Principally, it ensures that people are unprepared from a truth that is, in fact, frightening.
>
> *(Salamon 2020: 18)*

She goes on to convincingly argue that "claims that 'fear doesn't work' are not only patronizing and cynical; they have also been devastating in terms of mounting a real and timely response to the crisis" (Salamon 2020: 18). Furthermore, in terms of using comedy as a form of climate change communication in the manner in which Rod does, people often laugh at things that worry them, making humor a device for coping with the difficult issues. As a climate catastrophist, Rod is in good company. Frankel observes:

> In contrast to those socialists believing in the crisis-collapse of capitalism which would give rise to a socialist society, currently, there are many "catastrophists" or "collapsologists" who believe that there will be no post-capitalist society because "civilisation" itself will perish from climate breakdown. It is difficult to argue against "catastrophists" because climate breakdown could possibly engulf all types of political regimes rather than just advanced capitalist societies. Even those who do not subscribe to this view of climate catastrophe continue to express fears that social breakdown of capitalism – whether slow or rapid – will be characterised by the failure of new post-liberal institutions and social relations to triumph while the old order falls apart.
>
> *(Frankel 2020: 323)*

While like Rod, I have sometimes felt depressed about the socio-ecological crisis and catastrophic climate change, particularly in the wake of massive bushfires in southeastern Australia in late 2019 and early 2020 and more recently the COVID-19 pandemic, I generally resist such sentiments in the belief that one must struggle in the face of adversity, despite the odds.

I told Rod that there are efforts, including my own, to grapple with various visions that seek to transcend capitalism and replace it with a more socially just, democratic, and sustainable system, aside from whether it is called eco-socialism, eco-anarchism, Earth democracy, global democracy, or whatever. While initially reluctant to do so, I decided to prod Rod as my subject, comrade, and friend on his dystopian view of the fate of humanity, particularly due to its lack of significant political action on confronting climate change and to seriously reduce greenhouse gas emissions. In my Next System essay, which Rod and I discussed in our sixth interview, while not seeking to create a blueprint per se for creating an alternative world system that will be manifested in different ways in the many countries around the world, I propose several system-challenging reforms that potentially could facilitate a transition from the existing capitalist world system to a democratic eco-socialist world system (Baer 2016). In my essay, these include (1) the creation of new left parties designed to capture the state; (2) implementing greenhouse emissions at the sites of production that include measures to protect low-income people; emissions taxes at the sites of production; (3) increasing public ownership, socialization, or nationalization in various means of production; (4) expanding social equality within and between nation-states and achieving a sustainable global population size; (5) building workers' democracy; (6) creating meaningful work and shortening the work week; (7) achieving a net-zero-growth economy; (8) adopting energy efficiency, renewable energy sources, and green jobs; (9) expanding public transportation and massively diminishing reliance on motor vehicles and air travel; (10) resisting the culture of consumption and adopting sustainable and meaningful consumption; (11) introducing sustainable trade; and (12) building sustainable settlement patterns and local communities.

In my essay, I argue that as humanity proceeds into the 21st century, its survival as a species appears to be more and more precarious, a view that both Rod and I share, particularly given the impact of climate change in a multiplicity of ways looms on the horizon. I maintain that it is essential for critical scholars and social activists, including climate justice activists, to envision future scenarios and strategies for achieving an alternative world system. Perhaps more important is developing strategies to shift from the existing system of globalized capitalism to an alternative that transcends its numerous contradictions and limitations. While presently and for the foreseeable future, the notion that alternatives to global capitalism may be eventually implemented in any society, developed or developing, or in several linked societies may appear absurd, history tells us that social changes can occur very quickly once certain social structural and environmental conditions have reached a tipping point, a term that has become popular in climate science.

Rod responded to my vision of democratic eco-socialism in the following way:

> The world won't go there. There is no time to go there. The next 10–15 years look bad – things are breaking down. Donald Trump and Brexit are big

problems. Those things need to be addressed before we think about revolution. There is growing understanding that climate change is a problem, but the political process is so slow. The Liberals will fall apart for some time. Labor will not do anything and is not standing up to Adani, a proposed coal mining project in the Galilee Basin of Queensland.

Rod maintained Australia is heading into more heatwaves and a long-term drought. While he found the points in my essay to be "realistic and rational" and "would solve many problems," Rod asserted that the "system is so complex" that he can't see people in places like Yemen and Brazil "talking about these things given the immediacy of their problems." Like James Lovelock (2009), Rod saw no meaningful solutions to the climate crisis other than pockets of people living in "self-sufficient islands" or enclaves here and there. He argues that the "world is becoming like Brave New World. We have swapped our freedoms for security and the products of capitalist society, but those things will dry up." In terms of his climate catastrophism, Rod is in good company. For example, Jem Bendell (2018), an academic working on sustainability issues based at the University of Cumbia in the UK, penned an occasional paper titled "Deep Adaptation: A Map for Navigating Climate Tragedy," in which he criticizes sustainability studies for focusing on technological innovations designed to mitigate climate change, many of which he views as useless. Instead, he argues that humanity is doomed and encourages his readers to reassess their personal lives in the face of "an inevitable near-term social collapse due to climate change" (Bendell 2018).

When I asked Rod whether he feels that I am wasting my time in proposing an alternative to the capitalist world system, he answered that I am, a sense that I sometimes feel myself. While Rod feels hesitant to talk about climate change to his various audiences, he admitted that his performances provide him with a needed source of income and even the hope that someday "someone in the audience will jump up" and propose a solution. When I asked Rod what we should do under the circumstances, he proposed to merely do what we have been doing, he as a left-wing comedian and social activist and I as an anthropologist increasingly interested in developing a critical anthropology of the future (Baer and Singer 2018). To paraphrase Gramsci, it is a matter of "pessimism of the intellect, optimism of the will." When I said to Rod, "We have our work cut out for us," he replied, "We sure do. It is a big challenge." While he and I did not include a discussion of the COVID-19 pandemic over the course of our conversations because they occurred prior to this event, this new challenge, which is also related to the machinations of capitalism, illustrates the enormity of the socio-ecological crisis.

I tend to agree with Joshi (2020: 222) in asserting that that climate catastrophism or doomism constitutes a "significant threat to the voice of collective activism that has sprung up at the end of the 2010s." In response to critics that it is too late for a socio-ecological revolution that would pave the way to eco-socialism both globally and within specific countries, including Australia, Magdoff and Williams' remarks seem most apt:

> Too late for what? To struggle for a better world means taking the world as it is and working to transform it. Although the ecological and political conditions and trends are in many respects quite desperate, we are not condemned to continue degrading the environment or our social conditions. . . . A certain amount of global warming will continue regardless of what we do with all of its negative side effects. . . . However, we can stop the slide to an even more degraded Earth, poorer in species and in the health of remaining species. We can use the vast amount of available human and material resources to reorient the economy to benefit all people.
>
> *(Magdoff and Williams 2017: 287)*

Mainstream and radical perspectives on transitioning Australia to a more environmentally sustainable and climatically safe society

In this section, I explore several mainstream and radical perspectives on transitioning Australia to a more environmentally sustainable and climatically safe society. These include the mainstream perspective that generally is embodied under the notion of ecological modernization, eco-anarchism, the de-growth perspective, eco-feminism, and the Indigenous perspective.

Ecological modernization

Ecological modernization has become a virtually hegemonic stance that asserts that environmental sustainability and effective climate change can be implemented by adopting more energy-efficient, environmentally friendly, and low carbon-emitting energy sources and manufacturing processes. The "sunshine industries" that are touting energy conservation, efficiency, and renewable energy have "been working with some environmental groups to make the case for tough targets to stimulate markets from their products (for example groups like E7 and the European Wind Energy Association) (Newell and Paterson 2010: 42). Helm (2017: 6) boldly asserts that climate change will be a solvable problem "only with the march of new electric technologies," such as solar energy and electric car batteries.

Much of the climate movement is focused on moving beyond fossil fuels, a worthwhile endeavor. However, just as capitalism operated on other forms of energy prior to the Industrial Revolution, capitalism could theoretically operate on renewable sources of energy, a form of *green neoliberalism*, which will require enormous sources to develop and maintain. For example, the Koch brothers have become major investors in windfarm, solar energy, and biofuel projects. Subalterns around the world are increasingly having their land and labor expropriated by mining companies, including ones that are providing resources for renewable energy operations and supposedly green technologies, such as electric cars and autonomous vehicles (Arboleda 2020). Even though some variant of green capitalism might bring down greenhouse gas emissions to some degree, it would not address the social inequities, limited democracy, militarism, the threat of nuclear warfare, and global pandemics such as COVID-19 that are by-products of global capitalism.

In March 2017 an Australian billionaire, namely Mike Cannon Brookes (co-owner of software company Atlassian), and an American billionaire Elon Musk (Tesla founder) made a bet as to whether Tesla could fix the South Australia electricity grid within 100 days (Reucassel 2020: 86–88). As a result of this bet, Tesla developed a battery situated in Hornsdale which sold power to the grid when it was most needed. Since then other large batteries were built around Australia, including Gannawarra and Ballarat in Victoria and Lake Bonney in South Australia to give a boost to the electricity grid during periods of peak usage.

In South Australia, both ALP and Coalition governments have subsidized batteries for homes, with higher subsidies for concession cardholders. Reucassel reports:

> By 2020 . . . small batteries spread over South Australia's homes already had over half the storage capacity of the big Tesla battery in Hornsdale. This was a policy advanced by a Liberal government too. In South Australia, renewables are a bipartisan issue, with the Liberal government pledging to be net 100 per cent renewable by 2020.
>
> *(Reucassel 2020: 89)*

The Tesla battery based in Jamestown is designed to stabilize the South Australia electricity grid, "jumping in quickly to counteract disturbances in voltage and frequency" and also stores "electricity generated by the adjacent Hornsdale Wind Farm, helping to buffer its output that naturally varies with wind conditions" (ATA Energy Projects Team 2017: 9).

While the federal Coalition government is a laggard on renewable energy, South Australia obtains over 50 percent of its energy from renewable sources, Tasmania 90 percent, and the Australian Capital Territory or Canberra 100 percent (Reucassel 2020: 91). Victoria has implemented incentives for household solar and has invested in large-scale wind and solar projects. Daylesford and nearby Hepburn Springs in Victoria obtained council and state support to construct two wind turbines in their communities (Reucassel 2020: 216). Although being committed to new coal and coal seam gas developments, Queensland has taken some steps in developing household solar energy and large-scale renewable energy projects.

Ross Garnaut, a high-profile economist based at the University of Melbourne, exemplifies the hegemony of ecological modernization as a climate change mitigation strategy par excellence. He maintains that Australia has the potential of becoming an energy superpower due to its ready access to renewable energy sources, particularly solar, wind, and geothermal. Garnaut (2019: 11) boldly asserts: "The full emergence of Australia as energy superpower of the low-carbon world economy would encompass large-scale early-stage processing of Australian iron, aluminium and other minerals." He maintains that Australia could provide for not only its own energy needs with renewables but economically prosper by exporting renewable energy to its long-time trading partners of Japan, South Korea, the UK, and the European Union along with its newer trading partners, particularly China, but also Indonesia, India, and other Southeast Asian and South Asian countries. As a neo-classical economist, Garnaut frames his version of ecological modernization within a capitalist framework: He argues:

> Climate change will not be stopped by ending development. The challenge is to change the relationship between economic growth and emissions of the greenhouse gases that cause climate change.
>
> *(Garnaut 2019: 16)*

As a techno-optimist, Garnaut (2019: 17) firmly believes that the transition from fossil fuels to alternative forms of energy is "possible without sacrificing living standards in currently wealthy countries or disappointing hopes for improving conditions in the developing world." He maintains that battery and hydrogen electric-power vehicles will soon outcompete and thus replace motor vehicles powered by the international combustion engine (Garnaut 2019: 131). Garnaut laments that at the present time Australia has been a "global laggard" in the uptake of battery and hydrogen-electric vehicles, with EVs accounting for a mere 0.2 percent of Australian automobile sales, as opposed to 3.8 percent in the world and 5.8 percent in China (Garnaut 2019: 137).

Hydrogen power, which the Morrison Coalition government, is promoting potentially has its downsides. Hydrogen production methods can result in greenhouse emissions. While hydrogen does not emit CO_2 at the point of consumption in fuel cells, at which point the principal by-products are water vapor and small amounts of nitrogen oxides, some hydrogen production methods utilize fossil fuels, such as the production of hydrogen through coal gasification and steam methane reforming. Kaitsu, Swann, and Quickie report:

> The Latrobe Valley project plans to use brown coal to produce hydrogen through coal gasification. Hydrogen made via this method is known as "brown hydrogen". This method of producing hydrogen is highly inefficient and polluting. The pilot project in the Latrobe Valley estimates that it will produce over thirty times more carbon dioxide than hydrogen in weight.
>
> *(Kaitsu, Swann, and Quickie 2019: 8)*

Steam methane reforming utilizes natural gas or methane and produces "blue hydrogen," releasing methane in the process. Conversion, hydrogen production through electrolysis, is more expensive than the other two production processions, but results in zero-carbon hydrogen or "green hydrogen," with oxygen being the only by-product in the production process.

While Garnaut does not employ the term "green capitalism," his genre of green capitalism fails to address the treadmill of production and consumption that contributes to the depletion of natural resources and environmental degradation, including ultimately climate change. Ironically, in a book with the title *Reset*, he continues with his ongoing faith in techno-fixes and market mechanisms to shift Australia out of the economic doldrums created by the pandemic and the dangers of the climate crisis (Garnaut 2021).

Green capitalism tends either to be oblivious to social justice issues or at best to downplay them or pay them lip service. It is important to note that some components of ecological modernization have the potential to serve as important mitigation strategies. However, as anthropologist Al Hornborg persuasively argues:

> What ecological modernization has achieved is a neutralization of the formerly widespread intuition that industrial capitalism is at odds with the global ecology. . . . The discursive shift since the 1970s has been geared to disengaging concerns about the environment and development from the criticism of industrial capitalism as such. But the central question about capitalism should be the same as it was in the days of Marx: Is the growth of capital of benefit to everybody, or only to a few at the expense of others.
>
> *(Hornborg 2001: 25–26)*

Thus, ultimately, technological innovations that on the surface appear to be more environmentally sustainable and energy-efficient must be part and parcel of a shift to a steady-state or net zero-growth global economy if they are to circumvent the Jevons paradox or the rebound effect in which consumption tends to increase as products become more energy-efficient, a phenomenon that is consistent with the growth paradigm associated with global capitalism.

Whereas Garnaut adopts a hard or top-down version of ecological modernization, largely driven by the corporate sector but with government support, energy analyst Ketan Joshi (2020) espouses a softer and partially community-based version of ecological modernization. He acknowledges that wind and solar farms may intrude upon community landscapes, be they urban or rural, noting:

> These machines will be built, in most cases, close to where humans already live. There will be bigger solar farms. There will be large-scale battery storage facilities, and there will be many new transition lines that serve the increased integration of solar and wind power. Each of these technologies will create a unique collection of trip hazards that, if not pre-empted, will result in the slowing of decarbonisation. This is not an engineering challenge, nor is it a political and economic one. It is a challenge that involves the human heart and the human mind, and we have no choice but to understand it and solve it if climate action is to have any hope.
>
> *(Joshi 2020: 48)*

While many Australian households, businesses, and universities have adopted solar panels, large solar facilities are "relatively new in Australia, having reached around 3 percent of Australia's south-eastern grid," sometimes encountering community opposition to their share scale (Joshi 2020: 45). Joshi (2020: 48–49) argues that reducing greenhouse gas emissions and thereby mitigating climate change needs to be more participatory and socially equitable than it has in the past. Drawing upon community involvement in Denmark and Germany, in the form of the *Energiewende* in the latter country, he points to the Hepburn Wind Farm in Victoria as a community-owned operation which has become very popular with residents in the surrounding area (Joshi 2020: 73–74). Referring to state and territorial government plans to become anywhere from 50 to 100 percent renewable by 2030, Joshi (2020: 85) observes, "[I]t is tough to find traces of mentions of community ownership and community engagement within these state-placed plans (with the exception of Victoria and the Australian Capital Territory)."

Renewable energy generators will require equipment and buildings that will have to be produced by manufacturing processes that will require fossil fuels and mineral resources. While renewable energy sources, including solar, wind, and geothermal, have the potential of being part of the process of mitigating climate change, in and of themselves they are not a panacea. As Wall (2010: 11) observes, "Even a renewable energy capitalism would still tend to degrade the environment through commodification of nature." Ted Trainer (2007: 2), an Australian social scientist and eco-anarchist, acknowledges the superiority of renewable energy sources over fossil fuels but maintains that the "very high levels of production and consumption and therefore of energy use that we have in today's consumer-capitalist society cannot be sustained by renewable sources of energy." Furthermore, wind and solar electricity generation requires very large parcels of land, which Australia at the moment has, much of it consisting of land over which Indigenous people have native title rights. However, in some parts of the world, such as much of Europe and Japan, a transition to 100 percent solar energy might be difficult because of its land requirements.

Renewable energy and the digital economy are highly reliant in their present forms upon rare earth metals (Pitron 2020). Rare earth metals are found in terrestrial rocks in infinitesimal amounts and are a subset of some 30 raw materials often associated with more abundant metals, such as iron, copper, zinc, bauxite, and lead. Extracting rare metals entails a very energy-intensive and often labor-intensive process. For example, the production of one kilogram of vanadium requires 8.5 tons of rock to be purified, one kilogram requires 16 tons of rock, and one kilogram of gallium requires 50 tons of rock (Pitron 2020: 3). The extraction of rare metals also entails huge amounts of water in the purification process. The extraction of cobalt, which is used for lithium-ion batteries in electric cars, requires heavy physical labor with picks and shovels and results in the pollution of nearby streams in the Democratic Republic of Congo. The manufacture of laptop computers and mobile phones accounts for up to 19 percent of the global production of rare metals such as palladium and 23 percent of cobalt (Pitron 2020: 42). The mining of lithium, a white metal lying below the salt flats of Bolivia, Chile, and Argentina, and a component for batteries in electric cars, requires huge volumes of water, thus depriving local communities of an essential resource for their livelihood. As McIntosh (2020: 143) observes, the "shepherds of the Atacama Desert in Chile fear for their future as the battery boom opens up huge demand for the lithium mined there." Furthermore, the extraction of rare metals may result in an array of negative health consequences, such as children not developing teeth and even cancer (Pitron 2020: 27–28).

Pitron argues that scarcity of rare metals is a monumental issue that could even spark geopolitical conflict, particularly given that at the present time China monopolizes their production:

> On the one hand, advocates of the energy transition are adamant that we can draw infinitely on the inexhaustible sources of energy generated by the tides, the wind, and the sun to make our green technologies work. On the other hand, rare metals hunters warn that we could soon run out of several raw materials. Just as we have a list of threatened animal and plant species, we may soon have a red list of metals nearing depletion. At the current rate of production, we run the risk of exhausting the viable reserves of fifteen or so base and rare metals in under fifty years; we can expect the same for five additional metals (including currently abundant iron) before the end of the century.
>
> *(Pitron 2020: 160–161)*

Tim Flannery is an internationally renowned field zoologist, paleontologist, explorer, and conservationist based in Australia who also has embraced ecological modernization. He received much attention for his book *The Weather Makers: The History and Future Impact of Climate Change* (Flannery 2005). He has written several other popular books on climate change, the most recent one being *The Climate Cure: Solving the Climate Emergency in the Era of COVID-19* (Flannery 2021). Probably more than Ross Garnaut, Flannery is the most visible public intellectual in Australia touching on the climate change crisis not only in terms of his writing but in terms of presence in the mainstream media and on the lecture circuit, including at the University of Melbourne, where I have heard him speak several times. He served as the climate commissioner under both the Rudd and Gillard governments from 2007 to 2013. In 2009 Flannery (2021: 26) served as the chairperson of the Copenhagen Council, an "alliance of business, community and climate groups built over the three years prior to the United Nation Climate Conference in Copenhagen (COP 15) to assist the Danish Government (which was chairing COP15) to gain a successful outcome."

Whereas he believes that the Morrison government handled the 2019–2020 mega bushfire in a disastrous manner, he argues that by and large, it has handled the COVID-19 pandemic in an exemplary manner, thus allowing "Australia to mount one of the most effective COVID-19 responses to the initial round of infection in the world" (Flannery 2021: 9). He asserts that the federal government's response to the COVID-19 pandemic provides guidelines indicating how quickly Australia could address the climate emergency if the political will were present. Flannery delineates three critical battles required to address the climate crisis in Australia:

- Reducing fossil fuel use drastically and replacing it with renewable energy resources;
- Minimizing the damage that Australia and the planet will suffer as a result of greenhouse gas already emitted and those we cannot avoid emitting in the future;
- Creating the foundations of medium- to long-term responses to climate change (Flannery 2021).

In the event that the present Morrison government and future federal governments fail to adopt climate action policies, Flannery delineates various alternative strategies, namely:

- Getting state, territorial, and local council governments to set ambitious renewable energy targets;
- Organizing climate protest rallies operating in the spirit of groups such as Extinction Rebellion;
- Electing independents, such as Zali Stegall, who are willing to take climate action to the federal parliament;
- Becoming part of the divestment from fossil fuels movement spearheaded by American environmentalist Bill McKibben;
- Promoting hydrogen as one strategy for making Australia a clean-energy superpower along lines envisioned by Ross Garnaut;
- Promoting electric transportation, such as electric buses, trackless trams, electric cars, and possibly the long run electric aircraft over, by mid-century if ever;
- Developing hydrogen-powered ships;
- Developing a climate change adaptation plan for Australia formulated by an apolitical National Commission for Climate Adaptation;
- Drawdown or removal of carbon dioxide from the atmosphere by various methods (such as afforestation, covering nine percent of the oceans with seaweed, and various technological techniques), and storing it in ways that prevent it from returning to the atmosphere long-term (Flannery 2021: 79–180).

While space does not permit an elaborate discussion of the pros and cons of Flannery's alternative strategies, failing meaningful federal government climate action, his approach conforms to a green capitalist framework that fails to address social justice issues and the limits to growth.

Bearing this thought in mind, I now explore various radical perspectives, namely eco-anarchism, eco-feminism, the de-growth perspective, and the Indigenous perspective on the socio-ecological crisis from which eco-socialists can draw and hopefully vice versa, particularly in the Australian context.

Eco-anarchism

In terms of environmental issues, Anderson (2006: 256) asserts that "[a]narchism has a long history of environmentalism, from early anarchist thinkers such as Peter Kropotkin to the influential social ecologist Murray Bookchin (1991), who links the exploitation of nature to the exploitation of human beings," in much the same vein as eco-socialists. Greek eco-anarchist Takis Fotopoulos maintains that "modern hierarchical society," which for him includes both the capitalist market economy and "socialist" statism, is highly oriented toward economic growth, which has glaring environmental contradictions.

Ted Trainer is an Australian eco-anarchist, whom I visited briefly at the University of New South Wales, where he was an honorary fellow in late 2009 and with whom I have corresponded since then periodically. In the November 3, 17, and 24, 2020, issues of *Green Left* we engaged in a dialogue between eco-socialism and eco-anarchism. While I do not agree with Trainer on all issues, including the role of the state and social movements, in bringing about a socially just and environmentally sustainable society, his thinking has influenced me in various ways. He is a scholar whose work eco-socialists should consider. While Trainer has written several books, I summarize some of the principal points in his book titled *The Transition of a Sustainable and Just World*, in which he explores the following themes about capitalist society in general:

- It is highly unsustainable;
- It has a highly unjust global economy;
- It fosters over-consumption;
- Scarcity is the defining feature of the coming era;
- It cannot fix itself;
- It requires massive and radical changes;
- The alternative entails a "Simpler Way" (Trainer 2010: 2–12).

Trainer delineates the following five core principles of the Simpler Way:

- Material living standards, on the whole, must diminish;
- There is a need for small-scale, highly self-sufficient local economies and communities;
- Local communities based on cooperative and participatory principles control their own affairs, largely independent of international and global social structures;
- A radically different economic system needs to be developed, one that is socially controlled, oriented toward meeting needs as distinct from maximizing profits and not driven by market forces and a growth paradigm;

- These changes require a radical shift in values and worldview, especially away from competition, greed, and acquisitiveness (Trainer 2010: 6–13).

Trainer's first principle applies primarily to people who are relatively affluent in developed and developing countries and obviously the super-rich. It clearly does not apply to people who are abjectly poor in developing countries, particularly ones in Sub-Saharan Africa, or the indigent poor, particularly among ethnic minorities, and homeless people in developed countries, including Australia. Many of these people need to undergo "appropriate development," which for Trainer (2010: 129–130):

> should be about improving all aspects of society, including the quality of food and water and health services, the opportunities for leisure and cultural activity, the level of debate and discussion, the processes for government and administration, the moral standards, the geographic and aesthetic conditions in which people live, citizenship and social responsibility, openness and accountability, social cohesion, equity, concern for the less fortunate, the quality of life, security, the conditions of the poorest, and especially ecological sustainability.

Trainer (2010: 139) has high regard for indigenous and peasant lifestyles, which are "typically highly collectivist, and whose economies are governed by customs and traditions, not profit and gain."

Terry Leahy, another Australian eco-anarchist, envisions a "gift economy" in which:

> Producers in all sectors should decide what to produce and how to distribute it based on the needs of other groups. The status of the givers would depend on genuine needs being met by the gift. Production in each sector would be aware of their dependence on the services of other sectors.
>
> *(Leahy 2011: 114)*

He argues that "hybrid strategies" or practices employed by producers or community members could serve as transitions from the existing system to a gift economy. Actually existing examples of hybrids include support for environmental taxes, willingness to pay a higher price for more environmentally sustainable products, and volunteering to work in community gardens, which already exist in abundance in Australia (Leahy 2011: 122). Leahy (2011: 122) argues that a "revolution by stealth" might occur as the pervasiveness and influence of hybrid organizations come to dominate the economy, nationally, regionally, and internationally. Conversely, any gift economy would probably "have to achieve either a near universal spread or the military capacity to defend attacks from capitalist forces" (Leahy 2011: 125).

Eco-feminism

Eco-feminism emerged out of the feminist, peace, and environmental movements of the late 1970s and 1980s. It seeks to eradicate not only injustice against women and environment but all forms of social injustice. Carolyn Merchant (1996: 7) observes:

> Many ecofeminists advocate some form of an environmental ethic that deals with the twin oppressions of the domination of women and nature through an ethic of care and nurture that arises out of women's culturally constructed experiences.

Eco-feminism assumes various genres, including liberal eco-feminism, spiritual eco-feminism, and socialist eco-feminism (Merchant 2005: 200–211).

Liberal eco-feminists seek to liberate women within the contours of mainstream society while nurturing the environment. Spiritual eco-feminists focus on the female principle as cardinal to Mother Earth, who nurtures and sustains all human beings. For instance, Australian spiritual eco-feminist Freya Mathews (1991) draws inspiration from her claim that Indigenous people view the Earth as the great Mother and womb of life, which they view as sacred. Radical eco-feminists challenge the North/South divide associated with capitalist patriarchy, in which "nature is subordinated to man; woman to man; consumption; and the local to the global, and so on" (Shiva, Mies, and Salleh 2014: 5). They maintain that patriarchal societies are built on five interrelated pillars: sexism, racism, classicism, capitalism, and environmental degradation.

Socialist eco-feminism makes the category of reproduction as opposed to production central to the notion of achieving a socially just and environmentally sustainable world system. Ariel Salleh, an Australian sociologist, has served as a long-time proponent of socialist eco-feminism. She argues that eco-feminism reaches for an "Earth democracy" that "reframes environment and peace, gender, socialist, and postcolonial concerns beyond the single-issue approach fostered by bourgeois right and its institutions" (Salleh 1997: x). Salleh (1997: 12–13) maintains the ecological crisis stems from a "Eurocentric capitalist patriarchal culture built on the domination of nature, and domination of Women 'as nature'." Nature must benefit all human beings rather than just privileged ones as is the case under capitalist parameters. For Salleh (2009: 291), eco-feminism constitutes an "'environmentalism of the poor', and even in the global North, women, as a result of their regenerative labours, experience kinds of poverty and pain that are unknown to men."

De-growth

The de-growth perspective has developed more or less in tandem with eco-socialism since roughly the late 1970s or early 1980s. At times eco-socialism and de-growth have conversed with each other, but at others, they have talked past each other. For example, Serge Latouche, a French economist and anthropologist who popularized de-growth, drew upon the work of various radical thinkers, including French Marxist Andre Gorz (1980), who challenged the growth paradigm by arguing that perpetual growth was impossible (Kallis 2018: 6).

Vincent Leigey, a spokesperson for the French and international de-growth movement, and Anitra Nelson, an Australian social scientist who at times has referred to herself as an eco-socialist, state:

> Degrowth is an invitation to go on the inevitably long journey of the decolonisation of our growth imaginaries, moving from cultural awareness to a systemic and material transformation changing our everyday practices. Degrowth insists on the deconstruction and re-evaluation of beliefs within, and relations between, capitalism and productivism, consumerism and materialism, development and the quasi-religion of economism, science and technology.
>
> *(Liegey and Nelson 2021: 12)*

Anthropologist Jason Hickel (2020: 204) has pinned his hope for a pathway to a post-capitalist world in the de-growth paradigm which proposes "reducing the material and energy throughput of the economy to bring it back into balance with the living world, while distributing income and resources more fairly, liberating people from needless work, and investing in the public goods that people need to thrive." In order to achieve a global post-capitalist economy, he notes a need to democraticize the World Bank, the International Monetary Fund, and the World Trade Organization; a debt jubilee freeing poor countries to invest in needed services; ending corporate land grabs; distributing land back to small farmers; and reforming regimes that provide high-income countries an unfair advantage in the global agricultural industry. Other steps that Hickel suggests for moving humanity to an "ecological civilisation" include ending planned obsolescence; cutting advertising; shifting from ownership to usership; eradicating food waste; scaling down ecologically destructive industries; shortening the workweek; eliminating "unnecessary" jobs and shifting workers to jobs in renewable energy production, public services, maintenance, and so on; shortening the workweek; capping wage disparities; and expanding the commons. However, his book which appears to have drawn considerable attention suggests that he is not conversant with the burgeoning literature on eco-socialism.

I fear that the notion of de-growth is a Eurocentric one, applicable more to the Global North than the Global South. Obviously, there are large sectors of developed societies and smaller sectors of developing societies that need to undergo de-growth, but the abjectly poor of particularly the developing countries, along with homeless people or indigenous peoples living on reservations in North America and Australia, need to undergo some sort of growth or development in terms of access to nutritious food, decent housing and sanitation, health care, and education.

While most de-growth theorists would probably agree, my reading of the de-growth literature is that it does not emphasize sufficiently social justice issues.

Indigenous voices

Indigenous people created a sociocultural system in which they "understood that nature would generously provide for all of their material needs in return for minimal human effort" (Bodley 2017: 32). In order to survive in a wide array of eco-niches, indigenous peoples around the world have developed *traditional ecological knowledge* "based on empirical observation adapted to local conditions, that has allowed them to evolve as the environment has changed" (Servigne, Stevens, and Chapelle 2021: 87). Some Indigenous Australians have been involved in the Caring for Country movement, which seeks to preserve, protect, and rejuvenate traditional lands reclaimed through native title laws, many of which had been designated Indigenous Protected Areas (Altman and Kerins 2012).

While welcoming a shift from fossil fuels to renewable energy sources, renowned Indigenous Australian writer Tony Birch (2016: 359) calls for a shift in "our social and relationship with the planet required to live more equitably with each other and non-human species." He further asserts:

> In the future, ecological and environmental maintenance will require collective thinking, commitment and effort on a global scale. Solutions to climate change will remain elusive without such an undertaking. Solutions will not come from a reliance on government. In fact, progress on climate change will remain stifled if governments dominate discussion. . . . Through an intellectual, cultural and, for some, a spiritual attachment to place and country, we can have productive outcomes.
>
> *(Birch 2016: 376–377)*

Two Indigenous climate action groups are the Wangan and Jagalingou Family Council, representing the Traditional Owners of the land in Queensland's Galilee Basin, and Seed, an Indigenous youth group affiliated with the Australian Youth Climate Coalition. In terms of the proposed Adani Carmichael coal mining project, the Wangan and Jagalingou Family Council states:

> If the Carmichael mine were to proceed it would tear the heart out of the land. The scale of this mine means it would have devastating impacts on our native title, ancestral lands and waters, our totemic plants and animals, and our environmental and cultural heritage. It would pollute and drain billions of litres of groundwater, and obliterate important springs systems. It would potentially wipe out threatened and endangered species. It would literally leave a huge black hole, monumental in proportions, where there were once our homelands. These effects are irreversible. Our land will be "disappeared."
>
> *(Wangan and Jagalingou Family Council n.d.)*

Fortunately, the Wangan and Jagalingou community has gained support from various progressive groups, such as the Australian Youth Climate Coalition, Get Up! and Stop Adani!

As an Indigenous youth climate action group, Seed asserts:

> Our vision is for a just and sustainable future with strong cultures and communities, powered by renewable energy. Climate change is one of the greatest threats facing humanity, but we also know it is an opportunity to create a more just and sustainable world.
>
> *(Seed n.d.)*

In keeping with the Rights of Nature movement emanating from Ecuador and Bolivia, 16 Indigenous nations signed a treaty called the Union of Sovereign First Nations of the Northern Murray-Darling Basis which pledges:

> [T]he rights of Mother Earth are upheld by all Nations. . . . And we pledge our commitment to ensuring respect and preservation of her inalienable rights and all things natural. We acknowledge that these guarantees are the absolute inherent rights to the human condition.
>
> *(quoted in Maloney 2019: 23)*

Due to European invasion and colonization in Australia beginning in the late eighteenth century, Indigenous people lost access by and large to their commons. Due to social activism on the part of Indigenous people particularly beginning in the early 1970s, highlighted by the creation of the Tent Embassy in front of Parliament House in Canberra, Indigenous people under the leadership of Eddie Mabo won a victory in the 1992 Mabo case (*Mabo and others v. Queensland*), in which the High Court of Australia ruled that the "communal ownership of land of Aboriginal people in fact constituted a unique form of title that had existed prior to colonization" (Maddison 2019: 78). Corporate interests, particularly the mining and the pastoral industries, resisted, resulting in a series of complicated legal decisions. The Native Title Act 1993 "validated all existing freehold and leasehold land titles, thus assuring farming and mining companies that their businesses would not be taken from them under a *Mabo*-based land claim" (Burgmann 2003: 78). The Native Title Amendment Act 1998 "restricted the scope of native title claims and promoted the co-existence of native title with other uses, including through the establishment of Indigenous Land Use Agreements (ILUAs)" (Maddison 2019: 79).

Indigenous Australians have been divided on how to relate to Native Title and the mining industry. Indigenous political activist and historian Gary Foley (2007: 119) maintains that many Indigenous Australian activists regard Native Title legislation as a "sell-out" and an "absolute denial of Aboriginal sovereignty." Indigenous activist Patrick Dodson views Native Title as a Western strategy for limiting Indigenous demands (Maddison 2019: 80). Indigenous anthropologist Marcia Langton argues that Native Title claimants have gained some benefits, including the return of large parcels of land, financial payments, employment initiatives, and contract opportunities (Langton 2011; Langton and Mazel 2012: 28). She argues that Indigenous participation in the Australian mining industry is essential for guaranteeing Indigenous livelihoods into the future and has expressed strong contempt for environmentalists and leftists who have joined with some Indigenous people in opposing the mining sector (Langton 2013). Indeed, coal seam gas developments in New South Wales have resulted "in intense debate among Indigenous parties, with some vehemently opposed to them, and others such as the New South Wales Council (NSWALC) aiming to become actively involved in coal seam gas and oil extraction" (Mercer, Rijke, and Dressler 2014: 290). For the latter, coal seam gas development is viewed as providing opportunities for economic prosperity and authentic self-determination.

While indeed mining might provide some relatively well-paying jobs for some Indigenous people, it runs the danger of incorporating them into an industry that is a huge driver of deforestation, environmental degradation, land grabs resulting in the eviction of indigenous and peasant peoples from their homelands, and biodiversity collapse around the world (Arboleda 2020). Furthermore, as Loewenstein (2013: 81) observes, "[T]he history of Indigenous Australians is littered with companies that have pledged the world to traditional owners, but ended up giving very little."

Reportedly many Indigenous groups view successful Native Title claims as a beginning rather than an ending. Native Title has provided little opportunity for Indigenous Australians to control access to land or its use. Aside from its purported benefits, while recognizing that Native Title has resulted in over 20 percent of Australia theoretically falling under Indigenous ownership, as anthropologist Jon Altman (2009) argues, it does not provide property rights of commercially valuable resources, such as minerals and water. As Indigenous anthropologist Suzi Hutchings (2019: 206) astutely observes, "the colonial project continues under the guise of native title, as preexisting land title for Aboriginal people in Australia becomes reconstructed as a gift given to Aboriginal Australians by a magnanimous federal government." The Stop Adani! campaign has demonstrated the shortcomings of Native Title, evidenced when the Wangan and Jagalingou Council lost its challenge against Adani on July 12, 2019. In the end, meaningful land rights for Indigenous people would require a radical restructuring of the Australian political economy, perhaps one in which given the socio-ecological crisis faced not only by Indigenous people but the majority of Australians would entail a shift from the logic of late capitalism, often termed as neoliberalism, to one that based on eco-socialist principles that are committed to social parity and justice and environmental sustainability local, national, and global levels.

The role of anti-systemic movements and pre-figurative experiments in creating a socio-ecological revolution

Efforts to create socialism in Russia, China, and other developing countries came out of anti-systemic movements that resulted in new political regimes that in part created more egalitarian societies, but unfortunately not more democratic ones. As Silber (1994: 6) observes, "movements to replace capitalism with a more equitable (and yes – despite the Soviet debacle – a more efficient) social arrangement are sure to be permanent features of the world's political landscape so long as capitalism rules." While the Global Left briefly went into a period of dormancy during the late 1970s and early 1980s, anti-systemic movements representing a wide diversity of voice from indigenous peoples (e.g., the Zapatistas in Chiapas), landless people, peasants, and anti-corporate globalization activists (e.g., the anti-WTO demonstration in Seattle in 1999 and the World Social Forum) have been challenging the parameters of neoliberalism or late capitalism around the world. Furthermore, as I have already reported, various voices have been challenging the environmental unsustainability of capitalism and the anthropogenic climate change that it engenders. Powerful anti-systemic movements will be needed to counteract the resistance of the powers-that-be in the capitalist world system, including in Australia.

Whereas social movements, such as the environmental, peace, women's, community, and identity, including ethnic rights movements, in developed countries tend to be middle class in their composition, social movements in developing countries tend to be populist and working class. Australia has a rich history of progressive social movements, many of them with an anti-systemic tone, including the Aboriginal movement, the women's movement, the green or environmental movement, and the anti-globalization movement (Burgmann 2003). Stilwell provides a litany of progressive social movements in Australia which challenge global and Australian capitalism, including ones that manifest the following aims:

- Rejuvenating the labor movement;
- Developing a critique of neoliberalism or economic rationalism;
- Promoting social justice;
- Defending the public sector;
- Rejuvenating the notion of citizenship;
- Constructing alternative community structures;
- Promoting sustainable technology;
- Protecting the natural environment;
- Promoting reconciliation with Indigenous Australians;
- Promoting cosmopolitanism;
- Exploring new political structures (Stilwell 2000: 22).

Although Stilwell lists the labor movement first in his litany of potential anti-systemic social movements, particularly due to several decades of being subjected to neoliberalism, as Frankel (2020: 345) asserts, "trade unions are defensive organisations that can sometimes promote new social ideas, but they are ill-equipped to lead their members in a full-scale challenge to the social system as opposed to specific campaigns about wages, work conditions and other issues," and I might add climate change. Thus, in order to adopt an explicitly anti-capitalist stance, unions need to collaborate with other anti-systemic movements in challenging global capitalism and struggling for an alternative system based upon eco-socialist principles and objectives.

According to Hardt and Negri (2009: 94–95), "only movements from below" possess the "capacity to construct a consciousness of renewal and transformation" – one that "emerges from the working classes and multitudes that autonomously and creatively propose anti-modern and anticapitalist hopes and dream." A profound challenge has proven to be and will continue to be how to create cooperation among the variegated progressive social movements in challenging global capitalism and how to resolve tensions between the environmental and labor movements and how to create a democratic global governance regime when some groups emphasize local autonomy (Chase-Dunn 2005: 184).

Anarchists are more apt than socialists to use the concept of pre-figurative actions or communities, which entails "building the new world in the shell of the old" or "creating local institutions to meet communal needs" (Starr 2005: 120). In a somewhat different vein, Hammond (2012: 224) argues that many social movements "attempt to anticipate in their own social relations the future society they aim to create" in three principal ways, namely, "community, democracy, and decommodification." Australia has a wide array of pre-figurative social experiments, including the Save our Suburbs movement, the Simplicity movement, local farmers' markets, community gardens, the Transition Towns movement, the permaculture movement, and urban eco-communities (Alexander and Gleeson 2018; Cooper and Baer 2019). Unfortunately, as Frankel (2018: 67–68) observes, "Most of these local community alternative models are strong on how to organise small-scale activities but weak on detailed conceptions of how local communities connect to larger regional, national and international political, economic, legal and socio-cultural institutions."

Transitional steps to shift Australia toward eco-socialism

Frank Stilwell, a prominent political economist based at the University of Sydney, delineated steps in the creation of a "new socialism" in Australia, what he termed a Fourth Way, which included the following "radical reforms":

- More equitably shared labor which incorporates modern technology contributing to liberation from menial work;
- Lower unemployment;
- A blending of market mechanisms and planning which incorporates environmental sustainability;
- Fairer taxes;

- Widespread public ownership which provides space for small businesses;
- Public control of investment;
- Economic democracy;
- Distributional equity which limits the income accruing to the highest-paid workers in managerial positions to the lowest-paid workers to a ratio of roughly 3:1;
- Possibly some decentralization of state operations;
- More balanced urban and regional development;
- Sustainable environment;
- Promoting local and national economic developments as challenges to global capitalism (Stilwell 2000: 126–134).

More recently, David McKnight (2018) has called for the creation of an Australian progressive populism. His vision of Australian progressive populism includes the following dimensions, some of them pros and others cons:

- Promotion of greater social equality, entailing support for the needs of ordinary Australians as opposed to the wealthy elite;
- Restructuring and regulating the Australian economy rather than overthrowing capitalism;
- Solidarity by seeking to represent the whole of the people based on "universal supports like health, education, pensions and more" through a welfare state;
- Revitalizing democracy by encouraging popular participation;
- The notion that a democratic government should be "of the people, by the people and for the people;"
- No intention to replace the ALP or the Greens with a new political party;
- Formation of broad coalitions in challenging power;
- Taking mass action through a united front (McKnight 2018: 172–175).

While I am sympathetic with many of McKnight's objectives, I found them weak in terms of adequately addressing social justice and environmental issues, particularly the climate crisis, as well as resistance to the notion of transcending capitalism. In essence, how can you expect the system that created the problem to solve the problem?

In the remainder of this section, I draw up my earlier work on developing a pathway to eco-socialism at the global level in my Next System Project essay (Baer 2016) and my book titled *Democratic Eco-Socialism as Real Utopia* (Baer 2018) in an effort to develop a pathway to envision an eco-socialism in Australia. While I do not intend to present a blueprint as such for moving Australia toward eco-socialism, I seek to present rough guidelines designed to provoke discussion and break the impasse encountered by many eco-socialists on how to go from A to B, namely the existing capitalist world system to eco-socialism as a global phenomenon, with multiple national expressions, including in Australia. My guidelines, not necessarily in sequential order, are: (1) creating a new left party to capture the Australian state; (2) revisiting and expanding public ownership of the means of production; (3) implementing a steep emissions tax; (4) increasing social equality and creating green jobs; (5) shortening the workweek; (6) implementing workers' democracy; (7) challenging the culture of consumption and adopting a "Simpler Way"; (8) promoting sustainable settlement patterns, housing, transportation, and travel; (9) implementing sustainable agriculture and forestry; (10) implementing population and immigration policies that favor the most-needy people in developing countries; and (11) partially delinking from the capitalist world system. Needless, to say there is some overlap between the steps that Stilwell delineated two decades ago and my listing.

A new left party to capture the Australian state

Around the world, new left parties have emerged in recent decades. Some – namely, Bolivia, Ecuador, and Greece, albeit with mixed results for both internal and external reasons – have formed governments. Others, such as Podemos in Spain and DIE LINKE in Germany, have formed strong opposition groups within their legislatures or parliaments. Reflecting what's good about these efforts and learning from their failures, many of the other transitional radical steps delineated here would be much easier to implement if a new left party representing ordinary people rather than corporate interests and the rich and powerful would take the helm of the Australian state. This party might constitute an "unholy alliance" of various progressive groups and individuals – disaffected left-wing Australian Labor Party types, radical Greens, Socialist Alliance members, Solidarity members, Socialist Party of Australia members, various other socialist groups, and independent eco-socialists and even eco-anarchists. The notion of such a party might remind long-time Australian progressives of earlier unsuccessful initiatives.

In 1967 a "coalition of the left" convened a congress that reportedly "seemed to be a socialist government in the waiting at times" and constituted an "alliance of various classes and social forces whose objective interests were opposed to those of monopoly capital" (Murphy 1993: 231). Although this effort was led by the Communist Party of Australia (CPA), the CPA relinquished its traditional role of leadership in popular front campaigns and sought to build a "pluralist coalition around particular struggles" by recognizing the "legitimacy of other leftist, radical movements" (Murphy 1993: 231). Over the Easter break in Sydney in 1969, the CPA again convened a Left Action Conference which was attended by "some representatives of most of the New Left groups," including the Sydney-based Vietnam Action Campaign but not various Melbourne-based Maoist groups (Murphy 1993: 232).

Perhaps the best example of leftist unity in Australian history was the Moratorium campaign initiated in October 1969, which resulted in large demonstrations in various cities and towns on May 8, 1970, including reportedly 70,000–100,000 demonstrators in Melbourne alone (Murphy 1993: 241). While the Moratorium campaign was short-lived, it included the ALP Left, various peace and anti-conscription groups, some Old Left trade unions, various church groups, the New Left, an emerging women's movement, and a revitalized CPA (Murphy 1993: 249).

Other later efforts at leftist unity included the New Left Party, formed in the early 1990s, the Rainbow Alliance, which was active between 1987 and 1993, and, more recently, the Progressive Labor Party, which eventually imploded (Stilwell 2000: 165–170). The Rainbow Alliance (RA) formed in 1988 and had about 1,000 members, two-thirds of them situated in Victoria (Robinson 2019: 34). It constituted a "range of left activists, including many former ALP left members (but very few ex-Communists) such as Joseph Camilleri" and combined socialist and green agendas in a commitment to curtailing the capitalist exploitation of people and the environment (Robinson 2019: 34). RA supported candidates in two Victorian electorates and the candidacy of Jane Powell, the former leader of the Democrats, for the Senate in the federal election of 1993, but dissolved in 1998. In the case of the New Left Party, Robinson reports:

> In Australia, the crisis of socialism was most apparent in the collapse of the CPA. The Party dissolved in 1991 and its newspaper *Tribune* ceased publication the same year. The CPA was never an electoral force, but it occupied a central position in the culture of the left as an alternative focus of the ALP. It would take twenty years for the Greens to establish a similar alternative. Some members of the CPA tried to carry on its legacy free of the deadweight of Stalinism via the New Left Party (NLP), but this failed to attract support beyond former Communists and soon dissolved. *Broadside*, an attempted continuation of the *Tribune* with the ALP left participation, ceased publication in 1993.
>
> *(Robinson 2019: 2–3)*

Lessons learned from then would include how to build better coalitions, avoid personality conflicts and sectarianism, and develop democratic decision-making processes in challenging the capitalist hegemony worldwide and within Australia.

At this point in time, radicals in Australia have four strategic political options to move the country further to the left: (1) seeking to move the ALP further to left, even to move it to its social democratic form up to the 1970s; (2) seeking to move the Greens to the left; (3) being part of an alternative Coalition of sorts consisting primarily of the ALP and the Greens; and (4) forming an explicitly anti-capitalist New Left party.

Seeking to move the ALP to the left

At present, Australia's government is controlled by the Coalition, an alliance of the Liberal Party and the National Party. The Liberal Party constitutes the "main party of urban capital and big business" and has a long record of fostering privatization, at both the federal and state levels. In New South Wales, the Liberal government reportedly "sold off billions of dollars' worth of electricity, water and health infra-structure" (Denniss 2018: 17). The Victorian Liberal government (1992–1999) sold numerous state assets, including the State Electricity Corporation (SEC), largely to foreign corporate interests (North 2016: 72).

The National Party, formerly called the Country Party, "draws its core popu-lar support from the middle classes by appealing in particular to owners of family farms and small business in country towns" (Pietsch 2005: 28–29). However, as the case in the United States, farming has increasingly become the domain of huge agri-businesses. The Nationals also have close connections with the mining sector. Due to a feeling of disenfranchisement from the National Party, some middle-class people in regional cities and rural areas shifted their attention to Pauline Hanson's xenophobic One Nation Party beginning in the late 1990s.

Historically, the Australian Labor Party has been the working-class party and defined itself for a while as a "democratic socialist" party with Fabian or social dem-ocratic rather than Marxist roots. While it emerged in the late nineteenth century out of the Australian labor movement and has had a long-time alliance with trade unions, particularly since the early 1980s, this alliance has broken down as ALP politicians have become professionalized and have come to embrace "economic rationalism," the Australian term for neoliberalism (Bramble and Kuhn 2010). Whereas neoliberalism made headway under Margaret Thatcher in the UK begin-ning in 1979 and under Ronald Reagan in the United States beginning in 1981, in Australia it did so under two ALP prime ministers – Bob Hawke (1983–1991), former president of the Australian Council of Trade Unions, and Paul Keating (1991–1996). The adoption of neoliberalism by the Hawke and Keating govern-ments reduced public involvement in the economy by further lowering tariffs, introduction competition policy, floating the Australian dollar, and deregulating the banks. Australian neoliberalism continued its ascendancy during the Coalition government under Prime Minister Howard (1996–2007) and both Coalition and ALP governments since.

With some modification, Noam Chomsky's quip that the United States has a one-party system – the "Business Party" with two factions – applies to Australia. Fernandes (2018: 2–3) argues that whereas "power in Australia lies in the pri-vate sector where most important economic decisions are made," the state "has to facilitate corporate efforts to obtain revenue." Paul (2016: 2) maintains that what he terms "Australia's neoliberal corporate security states" constitutes a "hegemonic order backed by both mainstream political parties" under whose rule there has been a transfer of "considerable public power and wealth of corporations and wealthy and private interests" and a considerable increase of political and economic inequal-ity. Furthermore, he maintains that the Australian political economy functions as an "integral part of the American imperial project for a neoliberal free-trade global economy," one that makes Australia an US client state increasingly participating in US" (Paul 2016: 2).

Under the watch of the Hawke government, over $AU33 billion in public assets were privatized, including Australia's 22 largest airports, all public banks, large chunks of Australia's maritime and land transport infrastructure, the Commonwealth Serum Laboratories, and the Australian defense industry (Paul 2012: 12). By 2014, most major economic sectors in Australia were controlled by foreign corporations, with Anglo-American capital controlling over 50 percent of foreign investment. JP Morgan Nominees and Combined City Corps held large shareholdings in Australia's four big banks, namely Commonwealth, National Australia Bank, Westpac, and ANZ (Paul 2016: 13). In recent years, Chinese investment has penetrated the Australian political economy. China Light and Power, based in Hong Kong, owns Energy Australia, one of the big electricity providers with nearly 3 million customers in Australia's eastern states (Hamilton 2018: 121).

Between the 1940s and the 1960s, various Australian Trotskyist groups adopted the tactic of joining the ALP or *entrism* on the grounds that it constituted a "bourgeois workers party" and a mass party of the workers rather than simply a "bourgeois party" which could be moved leftward (Percy 2017: 62). Given the grim reality that the ALP has turned its back by and large on the Australian labor movement and the Australian working-class, many progressive Australians concluded some time ago that a "vehicle for radical reformism must be sought elsewhere" (Stilwell 2000: 137). On the matter of whether socialists should join the ALP in order to move it toward socialism, Armstrong and Bramble astutely argue:

> Starting from the assumption that Labor is the mass party of the working class, these socialists suggest that any attempt to build anything outside and against it is doomed to isolation and failure. This argument conveniently forgets that Labor is the political voice of the union *bureaucracy* and its middle-class parliamentary allies, not that of the working class. These forces will fight to the death any attempt to return Labor into a revolutionary organisation. Ultimately, the strategy of working within the ALP leads to either demoralisation or fatal political compromises on the road to winning positions and influences.
>
> *(Armstrong and Bramble 2007: 60)*

For many ALP stalwarts who lament the rightward movement of their party during the 1980s, Prime Minister Gough Whitlam (1972–1975) personifies the great reformer, whose short-lived government was overturned by the Coalition opposition leader Malcolm Fraser and Governor General John Kerr in the greatest political coup in Australian history. Indeed, the ALP Whitlam government introduced much-needed reforms, including substantial wage rises, a 36.25-hour workweek, paid maternity leave in the public service, equal pay laws for some women, national land rights laws, single parents' pensions, a nationalized health care system called Medibank, and free tertiary education (Armstrong and Bramble 2007: 73). In reality, Whitlam belonged to the ALP Right, ended the ALP's commitment to pull conscripts out of the Vietnam War, reinforced the US-Australia military alliance, and purged the left-wing Victorian ALP branch. In essence, as Bramble asserts, "Labor governments have never been interested in socialist transformation, for the simple reason that they are an integral part of the capitalist class rather than a threat to it." Nevertheless, the ALP depends heavily on the unions for financing but increasingly has sought funding from the corporate sector which in return lobbies it for support in a wide array of endeavors, including mining, energy development, property development, infrastructure projects, and the mass media (Bramble 2005: 81). Whereas in the past the ALP drew heavily upon blue-collar workers, its membership has become "dominated by people from professional occupations" (Bramble 2005: 83).

Can the Greens be moved further to the left?

As a consequence, many progressive Australians have looked at the Greens as an alternative for deeper systemic changes. A notable political difference between the United States and Australia is the presence of the Greens in the federal parliament – nine in the Senate and one in the House of Representatives, as well as many more in state parliaments and city councils. Thanks to a complicated system of proportional representation and preferential voting, several other minor parties also have seats, particularly in the federal Senate. However, what is desperately needed in the House of Representatives of the federal Parliament is the adoption of proportional representative along the lines of what presently exists in the Senate. Stilwell (2000: 175) asserts: "Electoral reform in the form of proportional representation is a key ingredient for getting a toehold into the parliamentary arena," perhaps along the lines of proportional representation in the German Bundestag and the New Zealand Parliament.

Brad Orgill (2013: 18) has called for another coalition of sorts in which the ALP and the Greens do not necessarily amalgamate but function as "long-term allies" in opposition to the Coalition Party. Indeed, given that the combined primary votes going to ALP and Green candidates in many federal elections often outnumber the primary votes going to the combined primary votes going to the Liberal and National Parties, an alliance of the ALP and the Greens could ensure its victory over that of the Liberal/National Coalition, thus ensuring its ability to take government. However, Orgill's call for an ALP/Greens alliance is still framed within capitalist parameters evidenced by his assertion that capitalism constitutes the "best system we have discovered to date" (Orgill 2013: 3). Furthermore, despite a temporary alliance between the ALP and the Greens in implementing an aborted Carbon Pricing Mechanism under the Gillard government, since then so much bad blood has developed between the two political parties that it is difficult to see the sort of coalition for which Orgill calls.

The Australian Greens, the largest minor party, by and large constitutes a green social democratic party, attracting over time many disaffected ALP supporters. They trace their roots to the United Tasmania Group, which formed in opposition to the impending damming of Lake Peddler in Tasmania in 1972. The first state Green Party formally registered in New South Wales, having "originated as a reading group within the ALP" and formed "partly as a response to the perceived decline of socialist influence in the ALP" (Hilier 2010: 8). The Australian Greens formed as a national confederation of state parties in 1992, but the creation of branch parties in all states and territories followed in 2003.

Over the past 15 years or so, the Greens have become a significant player in parliamentary politics. The party's initial electoral advance early in the 21st century was fueled by public concern about social justice and environmental issues that the ALP was no longer promoting. Unfortunately, by and large, the Greens have developed outside the labor movement and have found their strength largely in relatively affluent inner-city suburbs (Frankel 2001: 223). In the 2001 federal election, the Greens opposed Australian involvement in the Afghanistan war and mandatory detention of asylum seekers – and more than doubled their vote. Their longstanding commitment to peace and humanitarianism, economic equality, and social justice helped the Greens capture support from ALP voters who felt the Greens' program looked more like the old ALP's than the new ALP's did (Cahill and Brown 2008: 262).

Based on a platform of peace and non-violence, grassroots democracy, social and economic justice, and ecological sustainability, the Greens' campaigns and policy initiatives have from the start promoted public transportation, public health, public education, and the rights of welfare recipients and trade unionists while opposing privatization. The Greens oppose regressive taxation, like the General Services Tax, and demand much higher taxes on corporations, especially mining companies. Former Greens Party leader Bob Brown (2004: 70) stated: "Our aim must be to accommodate both social and environmental wellbeing." The Greens also oppose privatizing utilities, as a matter of social democratic principle but also because privatization encourages energy consumption. Put positively, the Greens call for a "comprehensive public transport system, with critical components publicly owned and controlled" – subsidized to discourage motor vehicle use (quoted in Matthews 2010: 12).

The Greens could theoretically constitute the starting point for a broad new left party in Australia, particularly as the contradictions of global capitalism become more apparent to more and more Australians as socio-economic, environmental, and climatic crises intensify at home and abroad. Paul (2016: 91) contends that the Greens constitute the only existing political party to contest Australia's "existing 'one-party' neoliberal state," but also "potentially could be co-opted by the neoliberal security state." In reality, the Greens are a broad church party; most of its members are green social democrats but a minority are "water melons" or eco-socialists and eco-anarchists (Fredman 2013). The mainstream of the Greens historically has had difficulty dealing with the presence of socialists within its ranks. A meeting of all registered Green parties in Sydney in August 1991 ruled that members of the Democratic Socialist Party "should not have voting rights as they were members of another political party" (Jackson 2016: 69). When the Australian Greens formed in August 1992, its constitution "included a proscription clause, restricting members of other parties from joining the Greens" (Jackson 2016: 70).

Some eco-socialists fluctuate between working within the bowels of the Greens and within socialist groups, particularly ones with eco-socialist stances, such as the Socialist Alliance and Solidarity. However, as discussed in Chapter 4, the Greens leadership has by and large been hostile to explicitly anti-capitalist factions with their party such as the Left Renewal. Particularly under the leadership of Di Natale, the Greens mainstreamed even more than they had in the past. Unfortunately, Left Renewal in New South Wales has run its course, having been successfully quashed by the national Greens leadership, and there is no sign of similar development in Green parties in other Australian states and territories. Certainly, the time has come for socialists and anarchists outside the Greens to enter into serious dialogue with Left Renewal. For the moment, Adam Bandt as the Greens leader, aside from his earlier flirtations with socialism and Marxism, since 2019 has opted to operate within capitalist parameters and sought to explore possibilities for a yet defined Green New Deal for Australia.

The need for socialist unity

To the left of the Greens is an Australian socialist movement. Highly fragmented and all too often sectarian, it comprises assorted small groups, with the Socialist Alternative and the Socialist Alliance being the largest, and with Solidarity, the Socialist Party of Australia, the Socialist Equality Party, the Socialist Freedom Party, and the Communist Party of Australia, being smaller and generally less visible publicly. Efforts to create socialist unity among several of these socialist groups have failed, particularly because they have different constituencies and partly because of doctrinal differences. The Socialist Alliance, the publisher of *Green Left*, has members serving in two city councils. The Socialist Party has a councilor in one city, while the Socialist Alternative is particularly popular among university students, who flock to its annual Marxism conference in Melbourne.

In the period 2012–2013, the two largest socialist groups, namely the Socialist Alternative and the Socialist Alliance, engaged in discussions about a possible fusion or merger of the two organizations which would seek to build a "mass revolutionary party which has its ultimate goal replacement of capitalism with a profoundly democratic, human and ecological sustainable system – socialism" (Fredman, Hinman, and Price 2013: 59). The Socialist Alliance emerged in 2001 out of an initiative on the part of two Trotskyist groups, the Democratic Socialist Party, formerly the Socialist Workers Party, and the International Socialist Organisation, the two largest socialist organizations in Australia. Various smaller socialist groups joined the Socialist Alliance. Corey Oakley, a stalwart member of the Socialist Alternative, states:

> Socialist Alternative joined initially, as we considered the aim of the Alliance worthwhile, but we were always sceptical about its prospects for success and our support was really only nominal. After about a year we considered that our scepticism had been vindicated and we left.
>
> *(Oakley 2013: 3)*

In time the various socialist groups that were part of the Socialist Alliance drifted away, with the Democratic Socialist Party keeping the Socialist Alliance alive as essentially a DSP front group. Given that this effort entailed a considerable amount of energy and expense, in January 2010 the DSP opted to collapse itself into the Socialist Alliance. The ISO, which had links with the Socialist Workers Party in Britain, joined with another small socialist group to become Solidarity, which has adopted, like the Socialist Alliance, a strong eco-socialist orientation.

Eventually, the effort on the part of the Socialist Alternative and Socialist Alliance of creating socialist unity through fusion once again failed when the Socialist Alternative withdrew from the negotiations process. Initially the Socialist Alternative, in contrast to the Socialist Alliance and Solidarity, opted to keep itself at a distance from the ecological and climate crises, arguing that it was too busy with other more immediate political campaigns. However, while not fully embracing eco-socialism per se, the Socialist Alternative at its annual Marxism conference in Melbourne, not convened in 2020 due to the COVID-19 pandemic, and in its two publications, a newspaper titled *Red Flag* and a journal titled *Marxist Left Review*, has become over the course of the past decade to grapple with the ecological and climate crises (see Ross 2011).

A more successful effort at socialist unity occurred in the guise of the Victorian Socialists in the November 24, 2017, Victorian state election. The newly minted party stood three candidates in the bid for the Upper House seat of Northern Metro. The number one candidate was Stephen Jolly, a former Socialist Party of Australia member and a Yarra city councillor of at the time 14 years; the number two candidate being Sue Bolton, a Socialist Alliance stalwart and councillor in the Moreland City Council; and the number 3 candidate being Colleen Bolger, a Socialist Alternative member and a lawyer representing people with asbestos-related diseases. In its manifesto, the Victorian Socialists campaigned for a jobs and industry plan, affordable housing, expansion of public transport, communities that could "resist racism and over-policing," better schools, more green space, enhanced union rights, public ownership of power and public transport, the creation of a Great Forest National Park, a "transition from old-growth logging to plantation timbers and replaceable products," a public investment program to transform the Latrobe Valley into a renewable energy hub, abolition of the scapegoating of migrants, abolition of discrimination and racism against Indigenous People, programs to address "gendered poverty and women's health and family, domestic and partner violence, and action on LGBTIQ discrimination (Victorian Socialists 2018: 8). In the federal election of March 2019, Victorian Socialist candidate did relatively well in three electorates, with Kath Larkin winning 4.7 percent of the primary votes in the electorate of Cooper, Sue Bolton 4.8 percent in the electorate of Wills, and Jerome Small 5.2 percent in the electorate of Calwell, well above socialist candidates in federal elections in many years. (Oakley 2019).

In reality, despite their Trotskyist roots, the Socialist Alliance, Solidarity, and the Socialist Alternative rarely these days invoke the ghost of Trotsky, although they are more likely to cite Marx and even Lenin, depending on the occasion. Frankel observes:

> [The] proliferation of Trotskyist sects and offshoots makes Monty Python's film *The Life of Brian* look like social realism. Their history has been animated by endless doctrinal splits and serious problems over the lack of internal party democracy. A combination of sharp critique and dogmatic "revolutionary purity" continues to guarantee their political irrelevancy.
>
> *(Frankel 2020: 205)*

While I would not say the various Australian socialist groups with a Trotskyist history are irrelevant, I would say as a member of the Socialist Alliance since 2009, that failure among them to achieve socialist unity is a tragedy. Groups such as the Socialist Alliance, Solidarity, and the Socialist Alternative may agree on 90–95 percent of the issues, but the 5–10 percent differences keep them at loggerheads with each other, on issues such as what it takes to be a "revolutionary" party or group or whether Cuba constitutes a revolutionary society or a state capitalist one. While Australian socialist groups have occasionally achieved some semblance of socialist unity briefly under umbrellas such as the Socialist Alliance and the Victorian Socialists, these alliances thus far have proven to be highly fragile. Nevertheless, I believe that achieving socialist unity in an era of socio-ecological crisis is highly imperative and theoretically could be an important step in creating a new left party in Australia.

Further thoughts about creating a New Left Party in Australia

Returning to the matter of forming an Australian new left party, Olin Erik Wright asks:

> Should anticapitalist, progressive activists work within established left and center-left parties, or form new parties? Should their efforts be concentrated at local, regional or national levels of contestation? What about movements and political parties? Because of the complexity and variability both in social context and political institutions, there can be no general formula to answer these questions.
>
> *(Wright 2019: 142)*

He suggests four guidelines for deciding whether to move forward with this complicated matter:

• Adherence to three clusters of values, namely equality/fairness, democracy/ freedom, and community/solidarity;

- Inclusion of explicit reform programs that connect identity-interests into the agenda of eroding capitalism;
- The incorporation of a deep or real democratic principles "in the interests of a very broad part of the population beyond the working class";
- A recognition that the "overall plan of eroding capitalism is not exclusively state-centered, and political parties are not the only collective actors needed for this strategy to be carried out" (Wright 2019: 142–143).

Given that the option of moving both the ALP and the Greens further to the left in the form of anti-capitalist parties appears unlikely, the only other option is the creation of an Australian New Left Party. Bearing in mind the principles delineated by Wright, this strikes me as the only path forward in terms of capturing the Australian state and being able to implement the other system-challenging or radical transitional steps that I have delineated. A new anti-capitalist left party holding the reins of state power could, in theory, implement many of the non-reformist reforms proposed in the following text.

Ultimately, efforts to create a new left party in Australia, like elsewhere, rely on mobilizing the grassroots, particularly anti-systemic movements. Such movements will pepper the whole global political landscape as long as capitalism remains a hegemonic political-economic system. Various anti-systemic movements – particularly, the labor, anti-corporate globalization or global justice, peace, environmental, climate, socialist, and anarchist, ethnic and indigenous rights, and women's movement – have important roles to play in creating a socio-ecological revolution committed to social justice, democratic processes, environmental sustainability, and a safe climate. These movements already interact, with mixed results, but they need to do much more to identify their unitary purpose, namely being part of an effort to transcend capitalism both globally and within Australia.

Revisiting and expanding public ownership in Australia

Revisiting social or public ownership of the means of production or restoration of common ownership of the land and of natural resources, something that Indigenous peoples practiced prior to European colonization of Australia, constitutes a vital climate change mitigation strategy. As Whyte so eloquently argues:

> The restoration of common ownership is necessary to remove the logic of capital accumulation from the stewardship of the natural environment. Only through a restoration of the commons can we displace the social dominance of capital and the corporation over the future of the planet and allow the damage to our environment to be repaired.
>
> *(Whyte 2020: 174)*

Even in the United States, public control of social services exists in generally conservative states, the so-called red states in terms of being Republican strongholds, such as Nebraska, which enforces public control over its electricity utility companies, and North Dakota, which has a state-owned bank (Kolaski 2021: 23). Historically, public ownership of productive forces in Australia has not been a particularly radical idea. Robert Menzies, the Coalition prime minister (1939–1941 and 1949–1961), stated as far back as 1943: "Few people would have any quarrel with government control of railways, or tramways, or water supply, or such other great public utilities" (quoted in Wettenhall 1965: 428). In 1912, the Fisher ALP government created the government-owned Commonwealth Bank of Australia to serve as both a Central Bank and a savings bank. The Tasmanian Hydro Department began operations as a state-owned operation in 1914. The Victorian State Electricity Commission was created in 1919 to electrify suburban railways and generate power from the brown coal fields in Gippsland. Numerous municipalities in Queensland generated and distributed electricity. Throughout Australia, many gas and water utilities were public-owned enterprises. As Lowe (2016: 136) observes, "[a]lthough Australia was governed from 1949 until 1972 by coalitions dominated by the Liberal Party, there was a national consensus that the government had a vital role in funding the essential infrastructure: roads, railways, airports, water supply, sewage systems and so on." He reports:

> In 1964 Australia produced radios and television sets, electronic equipment for communications and navigation, cars and larger transport vehicles, boats and light aircraft. Over the last 50 years, we have systematically run down our manufacturing capacity. The last large-scale vehicle building plants will close in the next few years. Appliances and electronic equipment in our stores come from overseas. We even import such basic products as clothes and shoes.
>
> *(Lowe 2016: 65)*

In the early 1970s, the short-lived ALP government extended this funding, covering a universal health care program called Medibank and free tertiary education. A Coalition government then essentially curtailed Medibank, but in 1984, the ALP-led government reintroduced a comprehensive national health plan called Medicare, which is still in place. While Medicare has its shortcomings and private health insurance plans have made inroads, here Australia stands in stark contrast to the United States – the world's only developed society without some sort of comprehensive national health plan.

Since the era of public ownership and movement toward universal healthcare and free tertiary education, Australia has moved in the wrong direction. Starting with the Hawke government,

> the power of big business to move its capital offshore was radically opened up. In December 1983, the Australian dollar was floated, government control over banks reduced, and foreign banks allowed to set up operations. The

Hawke government also gave open slather to the media barons, with the relaxation of the rules on the ownership of print and electronic media.

(Armstrong and Bramble 2007: 80)

Under the dictates of neoliberalism, such publicly owned enterprises as Qantas Airlines, Commonwealth Bank, Commonwealth Scientific Laboratory, Australia Post, and Telstra (a telecommunication firm) were privatized. Meanwhile, state governments have "sold most of the public utilities they once owned to meet the cost of building roads, schools and hospitals to sustain the rising population" (North 2016: 176). Free tertiary education ended in the late 1980s under the ALP government. As Davis observes,

> Every time something is privatised and taken out of public ownership, for example, wealth moves upwards because the privatisation of public assets shifts them into private hands. The things that were once owned by everyone, and which generated prosperity for everyone, are now owned by and generate profits for a privileged few – they are run in the interests of shareholders, most of whom are corporate investors. When corporations use up public resources to generate profits and create environmental costs, this, too, moves wealth upwards.

> *(Davis 2008: 101)*

Public ownership of numerous sectors of the Australian economy that have been privatized needs to be re-socialized. Such a development, difficult by any reckoning, will require a progressive government with the political will to carry it out and abide by the wishes of many Australians rather than those of multinational corporations with operations in Australia. Given the extremely high cost of housing, both purchased or rented, Australia also needs to shift away from private housing ownership to public housing, another hard slog where private homeownership is highly valued and where many people purchase second housing units as investments, further driving up the cost of housing and making it unaffordable for many young people looking to buy or rent.

Potentially implementing public ownership of the mining industry, particularly the coal mining industry, might be a starting point for introducing more public ownership in general and could serve as a necessary step toward the goal of creating a genuine an eco-socialist society in Australia. Such a move would reduce the power of the mining sector and wealthy individuals to influence elections through campaign contributions, favorable media coverage, and even bribery. Socialization or public ownership, both at the state level and in the form of cooperatives, could also facilitate the shift from fossil fuel energy sources (with their taxpayer funded-subsidies) to renewable energy and trigger what has come to be known as a "just transition" for displaced workers by, for instance, offering former coal miners jobs in solar panel construction and the development of wind farms and solar power plants. Climate change mitigation strategies to increase energy efficiency and discourage energy consumption would also be much easier to instigate if the energy supply was publicly owned.

While privatization's downside has become an axiomatic discourse among even relatively progressive actors, including the Greens, the discussion of ways to socialize energy production and shift away from coal production has larger dimensions. The shift must emphasize benefiting ordinary people rather than corporations, democratic rather than authoritarian structures and means, and the growth of human potential over economic growth.

Fortunately, movement is already underway. The Socialist Alliance has called for socializing or nationalizing many sectors of the Australian economy in its 13-Point Climate Action Plan. Climate action groups, environmental NGOs, and such energy think tanks as Beyond Zero Emissions, as well as the Australian Greens, all need to come to terms with the need to end corporate dominance of the mining industry/state nexus, through nationalizing or socializing energy production and generation, particularly of renewables such as solar, wind, and possibly geothermal energy. As the adverse impacts of climate change on Australia and the world become more apparent, this will get easier, though we can't wait for hell and high water to get the point widely accepted.

Making the transition harder to sell and implement, however, is the internationalization of energy ownership. Partly spurred by the devaluation of the Australian dollar, many large Australian-owned companies, such as BHP and Rio Tinto, have been over the past several decades taken over by foreign-owned corporations, thus making Australia even more of a semi-peripheral country and less of a core country in control of its own resources (Frankel 2001: 23). Socializing the means of production would allow for a return to a stronger domestic manufacturing sector.

Without seeking to romanticize the past, until around the 1960s, Australia manufactured many of its own consumer goods, whether these were automobiles, bicycles, appliances, or clothing apparel. Due to corporate globalization ever on the lookout for cheaper labor, many of these products are now manufactured abroad and then transported to Australia by ships or airplanes, resulting in greenhouse gas emissions. For example, over 80 percent of clothing for sale in Australia is reportedly imported from China (Green and Minchin 2010: 25). Nationalization of production could be part and parcel over capitalist-driven inbuilt obsolescence and contribute to the production of fewer but durable things.

On the international front, in 2002 the Coalition government under Prime Minister John Howard entered into the Australia United States Free Trade Agreement (AUSFTA) with the United States, a move that ALP supported. As with most free trade agreements, particularly ones with the United States, AUSFTA is very one-sided. Tariffs on US agricultural exports to Australia were eliminated while the United States hung onto many of its tariffs and quotas on Australian agricultural products. In a similar vein, the United States obtained nearly total access to Australian markets for manufactured goods but preserved trade barriers to Australian manufactured products, including Australian fast ferries. Australia is also a signatory of the US-initiated Trade in Services Agreement. Ironically, while the United States initiated the Transpacific Partnership (TPP), only to have former resident Donald Trump withdraw it from TPP negotiations upon taking office in 2017, Australia remains an intended signatory. As North (2016: 318) observes, "In Australia, the Productivity Commission, dominated by economists and obsessed by short-term prices of products, recommends that we drop tariffs even though doing so risks further sections of the manufacturing industry going to the wall."

A steep emissions tax for Australia

An emissions tax or carbon tax is fraught with danger because corporations typically find loopholes and pass the burden of the tax on to working-class people, compounding the primary injustice of climate change with secondary injustices. However, a carbon tax designed to prevent such unfair outcomes would help curb emissions. Climate scientist James Hansen and Yale economist William Nordhaus maintain that carbon taxation is crucial to mitigating climate change (Brohe, Eyre, and Howarth 2009: 155). This is especially so if the tax is extracted at the point of production. Hansen and others have proposed a "carbon tax to be implemented at the wellhead, mine gate or port of entry," given the flaws in other market mechanisms, particularly emissions trading schemes and carbon offsetting (Frank 2009: 36).

To prevent secondary injustices, carbon taxation must incorporate measures to ensure that poor consumers do not suffer. Stilwell (2000: 225) maintains that a carbon tax "would be both an effective revenue raiser and a disincentive to persist with the use of goods and services which are environmentally degrading." Dick Nichols (2010: 11), a Socialist Alliance member, compiles additional reasons for the tax: a rising carbon price is a "powerful incentive for capitalists to move to less carbon-intensive production methods"; a "properly designed carbon tax would make generous compensation schemes for working people and people on welfare faced with rising fuel and electricity prices"; and, if combined with the elimination of the General Sales Tax (GST), increases in welfare payments and cuts to income tax for the low-paid, it could provide a "tax mix" that "puts people and the environment before polluting corporations." More recently, Pittifor argues:

> Carbon taxes generate revenue and may be necessary to change behaviour, but like flat taxes, they are largely regressive. They could have an important but limited role to play as part of a far more fundamental reshaping of the economy, but to be effective they should be targeted at the biggest emitters.
>
> *(Pitron 2020: 105)*

Lin and Li (2011) found that the carbon taxes operating in Denmark, Norway, Sweden, and the Netherlands did not effectively reduce emissions because these countries provided exemptions to energy-intensive industries.

Of course, a carbon tax won't necessarily work if it is poorly designed. Australians saw that when former Julia Gillard's ALP government instituted a tax in 2012, that fell on the twin swords of administrative complexity and short-horizon politics. Intended to serve as an interim measure that would transition into an emissions-trading scheme after three years, the Carbon Pricing Mechanism (CPM) was crafted by the Multi-Party Climate Committee (ALP and Greens) to cut Australia's CO_2 emissions by 5 percent, based on 2000 levels, by 2020, and by 80 percent cut of those levels by 2050. Some 500 big polluters were to be taxed $23 for every ton of CO_2 emissions they produced, rising to $24.15 in 2013–2014 and $25.40 in 2014–2015. Agricultural emissions were exempted. The short-lived CPM included numerous subsidies for industries and low- and middle-income households and free permits for some industries. To encourage renewable energy, a Clean Energy Finance Corporation was created under CPM to invest in renewable energy projects and finance research and development of renewable energy. But after a complicated political coup in 2013, the new Rudd government whittled the now unpopular, poorly constructed, and poorly understood carbon tax from $23 per ton of CO_2 emissions to the then current EU carbon price of $9 per ton of CO_2 emissions. Then, in the crowning blow, the promise to scratch the tax altogether was part of Tony Abbott's winning bid in the 2013 election, and a year later it was gone.

What went wrong? During its brief lifespan, the CPM reportedly resulted in a modest 1 per cent fall in national emissions. However, electricity prices skyrocketed, not so much because of the carbon tax but because electricity companies had invested heavily in infrastructure, the costs of which they passed onto their customers. However, the Coalition opposition blamed the rise of electricity costs on the CPM, an assertion that many Australians accepted.

Increasing social equality and creating green jobs

While social inequality in Australia is not as great as it is in the United States, it has one of the highest degrees of social inequality of OECD countries. The top ten richest people in Australia in 2019 and their business backgrounds are depicted in Table 6.1.

Table 6.2 depicts the top 20 CEO "realized" pay packets for their company's 2019 financial year.

As in most countries, the gap between salaries and salary packages between CEOs and ordinary workers has widened over the past several decades. North reports:

> In Australia in 2014 the ratio of CEO average pay to worker average pay was 93, with the gap also widening each year. The average annual hike in executive compensation awarded to executives by themselves was about 20% per annum, largely regardless of performance.
>
> *(North 2016: 139)*

TABLE 6.1 Ten Richest Persons in Australia, 2019

Person	Businesses	Wealth ($US)
Gina Rinehart	mining, cattle, media	$14.8 billion
Harry Triguboff	real estate	$9 billion
Anthony Pratt	packaging and recycling	$6.8 billion
Frank Lowy	managing family owned investment house	$6.5 billion
Scott Farquhar	Software	$6.4 billion
Mike Cannon-Brookes	Software	$6.4 billion
Andrew Forrest	iron ore mining	$4.3 billion
John Gandel	shopping malls	$4.0 billion
James Packer	gambling casinos in Australia and UK	$3.6 billion
Lindsay Fox	logistics – business park, airports, and currency management	$3.5 billion

Source: Peterson (2020).

TABLE 6.2 Top 20 CEO "Realized" Pay Packets in Financial Year 2019

Rank	Chief executive	Company	Realized pay
1	Andrew Barkla	IDP Education Ltd	$37,761,322
2	Paul Perreault	CSL Limited	$30,526,634
3	Philippe Wolgen	Clinuvel Pharmaceuticals Limited	$20,624,450
4	Michael Clarke	Treasury Wine Estates Ltd	$19,853,177
5	John Guscic	Webjet Limited	$16,498,937
6	Greg Goodman	Goodman Group	$14,967,391
7	Robert Kelly	Steadfast Group Ltd	$14,419,677
8	Alan Joyce	Qantas Airways Limited	$12,217,400
9	Colin Goldschmidt	Sonic Healthcare Limited	$11,912,450
10	JS Jacques	Rio Tinto Limited	$10,323,975
11	Peter Coleman	Woodside Petroleum Limited	$9,665,221
12	Mark Vassella	BlueScope Steel Limited	$9.465,692
13	Mark McInnes	Premier Investments Limited	$9,155,382
14	Bill Beament	Northern Star Resources Ltd	$8,858,086
15	Julian Pemberton	NRW Holdings Limited	$8,815,450
16	Nigel Garrard	Orora Ltd	$8,595,076
17	Maurice James	Qube Holdings Ltd	$7,735,816
18	Paul Flynn	Whitehaven Coal Ltd	$7,619,735
19	Scott Charlton	Transurban Group	$7,609,185
20	Peter Allen	Scentre Group	$7,452,446

Source: Adapted from Yoo (2020) and data obtained from the Australian Council of Superannuation Investors.

An Australia Institute (2018: 1–2) poll surveying 1,557 people "online through Research Now with nationally representative samples by gender, age, and state and territory" indicated a large percentage disapproved of "very high executive salaries." On the item stating that a "strict limit on the maximum salary of a company can pay a CEO or executive staff member," 42 percent strongly supported this measure and 33 percent supported it (Australia Institute 2018: 1). On the question, "What do you think is a reasonable salary for a CEO at a large corporation?," 56 percent of the respondents ticked the item "720,000 a year or less," 73 percent ticked the item "1.8 million a year or less," and 80 percent ticked the item "5.4 million a year or less" (Australia Institute 2018: 9). On the item calling for a "strict limit on the maximum limit salary a company can pay a CEO or executive staff member," 72 percent of Coalition voters totally agreed, 77 percent of ALP voters totally agreed, and 74 percent of Green voters totally agreed (Australia Institute 2018: 7).

In terms of income, the gap between the top and the bottom in Australia has been increasing in recent decades. Table 6.3 depicts the share of income by deciles for between 1995–1996 and 2015–2016.

Goldie and Saunders (2018) conducted a more extensive statistical study in which they reported on both wealth and income inequality in Australia in 2016 based upon Credit Suisse Research Institute data. Table 6.4 depicts wealth income in $USD.

Table 6.5 depicts household incomes annually and weekly at extremes of the top and bottom of the income ladder

In 2015 or thereabouts, Australia had a level of income inequality or Gini coefficient of 0.337, higher than the OECD average of 0.317 (Goldie and Saunders 2018: 31). OECD countries with higher Gini coefficients than Australia in rank order starting from highest included Mexico, Turkey, the United States, Great Britain, and Greece. OECD countries with lower Gini coefficients than Australia in

TABLE 6.3 Income Share by Deciles in Australia

Fiscal Year	Bottom 10%	20%	30%	40%	50%	60%	70%	80%	90%	Top 10%
1995–1996	3%	4%	5%	7%	8%	9%	11	12%	15%	26%
2015–2016	3%	4%	5%	6%	7%	8%	10%	11%	14%	31%

Source: Adapted from Grudnoff (2018: 6).

TABLE 6.4 Wealth Inequality in Australia, 2016

	Highest 5%	Highest 20%	4th 20%	Middle 20%	2nd 20%	Lowest 20%
Wealth	$6,044,000	$4,217,000	$2,909,000	$964,000	$538,000	$237,000

Source: Adapted from Goldie and Saunders (2018: 20).

TABLE 6.5 Household Incomes – Average Annual Gross (Pre-Tax) and Average Weekly Disposable (After Tax Income) in 2015–2016 in USD

	Highest 1%	Highest 5%	Highest 20%	Lowest 20%	Lowest 5%
Annual Income	$950,844	$458,727	$354,362	$39,727	$23,265

Source: Adapted from Goldie and Saunders (2018: 27)

TABLE 6.6 Green Jobs Typology

	Deep Green	*Mid Green*	*Light Green*
Mode	Proactive	Integrative	Reactive
Scope	Long Term	Intermediate Term	Short Term
Nature	Transforming	Reforming	Conforming
Objective	Redefine Growth	Ecologize Growth	Enhance Growth
Operation	Rejectionist	Reinventionist	Accommodationist
Aim	Ecological Sustainability	Ecological Modernity	Sustainable Development
Jobs	Preserving Nature	Greening Industry	Remedying Ecological Decline

Source: Crowley (1999: 1017).

rank order starting from lowest include Denmark, Finland, Norway, Austria, Sweden, Germany, France, South Korea, Canada, and Japan.

Among developed countries, Australia is unique in having no tax on wealth or inheritance. Relinquishing that uniqueness for the sake of redistributing the wealth in Australia and providing state funds for social and health services would be an important step toward eco-socialism. But imposing a high inheritance tax on the affluent, or even creating a progressive tax structure with no loopholes, requires considerable political will. The same goes for a high corporate tax and a guaranteed livable income. While ALP's left faction periodically laments the mal-distribution of wealth in Australia, but the ALP as a whole in recent decades has not taken advantage of its stints in power to challenge it.

The notion of *green jobs* is a contested space that does not allow for a detailed discussion of the debate on this issue. However, from an eco-socialist perspective, I find Kate Crowley typology of green jobs depicted in Table 6.6 quite useful.

Light-green jobs conform by and large with "business as usual; mid-green jobs focus on integrating environmental concerns into existing industries; and deep-green jobs entail adopting renewable energy technologies and challenging the capitalist growth paradigm. However, it is important to note that a decarbonization of energy production does not simply translate into new long-term jobs in the renewable energy sector given that wind and solar farms require very few workers given they are highly automated operations (Warren 2019: 98).

Shortening the workweek

Even though Australian workers pioneered the eight-hour workday in the mid-nineteenth century, which eventually led to the 40-hour workweek, most full-time employed Australians now work over 40 hours a week, often well over. Indeed, overwork is rampant in the land down under. Whereas Australia once was viewed as a "workers' paradise," which led the world in paid holidays, paid sick

TABLE 6.7 Average Annual Labor Hours Per Worker in Selected OECD Countries, 2000–2016

Country	2000	2005	2010	2016
Australia	1,779	1,729	1,692	1,669
Canada	1,779	1,747	1,703	1,703
Denmark	1,466	1,451	1,422	1,410
Germany	1,452	1,411	1,390	1,363
Greece	2,108	2,136	2,020	2.035
Japan	1,821	1,775	1,733	1,713
Mexico	2,311	2,290	2,254	2,255
Sweden	1,642	1,605	1,635	1,621
United Kingdom	1,700	1,673	1,650	1,676
United States	1,834	1,795	1,714	1,783

Source: Adapted from OECD (2017).

leave, and the eight-hour work week, Australians "now work some of the longest hours in the developed world," with most workers working unpaid overtime and receiving paltry compensation if unemployed (Denniss 2018: 3). Overconsumption invites overwork for some, but for others at the lower end of the labor market, decreased pay associated with decreased unionization rates are at fault. Internationally, shorter working hours correlated with strong trade unions – and Australia has been drifting toward the US pattern of longer working hours and weaker trade unions. Table 6.7 depicts the average annual labor per worker in selected OECD between 2000 and 2016.

Shortening the workweek and bringing back the long weekend would reduce transport use to the extent that both moves reduced trips to and from workplaces. It might take the culture of consumption down a notch too, and both these checks on excess would help with climate change mitigation. The move would also challenge the culture of consumption by inviting people to spend time pursuing things besides money, possibly including time to grow more of their own food at home or in community gardens – another plus for climate stability. And, finally, if unions embraced this challenge and mobilized to reduce working hours, with increased hourly rates of pay as needed, they would also continue in their historic role of improving workers' well-being.

Workers' democracy

Australia has a rich labor history in which a relatively high percentage of wage workers belonged to a wide array of trade unions. Indeed, the labor movement and the ALP were much more closely intertwined than is the case today. In 1914, union membership in Australia stood at one-third of all wage-earners, a "level unprecedented in any other country" at the time (Macintyre 2016: 131). Union density in Australia peaked at 51 percent in 1976 but began to decline rapidly with the advent of neoliberalism in the early 1980s, standing at around 14 percent in 2018 (Gilfillian and McGann 2018). In 2018, as opposed to a trade union density of 91.8 percent in Iceland, 66.5 percent in Denmark, 49.2 percent in Norway, 25.9 percent in Canada, Australia's union density of 13.7 percent was roughly in the same ballpark as the union densities in Japan (17.0 percent), Germany (16.5 percent), and the United States (10.1 percent) (OECD 2021).

Despite the overall decline in union membership in Australia, unions are not disappearing and health, community services, and construction-related unions have grown in recent years. For example, the Australian Nursing and Midwifery Federation, the Australian Education Union, and the National Tertiary Education Union are holding their own under potentially crippling pressures from management. A first step toward rejuvenating Australia's union movement might be to loosen its ties with the ALP and help create a new left party that would foster workers' democracy.

And what would a workers' democracy look like? In publicly owned companies, workers and consumers would join in decision-making as part of the process of creating a "participatory economy" in an "authentically democratic society" (Dominick 2008: 385). Unfortunately, such enterprises are still very rare in Australia and Australian trade unions so far haven't even broached the possibility of workers' democracy in any form. An exception, still in its infancy, is the Earthworker Cooperative, formed to create a network of worker-owned cooperatives throughout Australia. It maintains that "social and environmental exploitation are intertwined, and that the problems of climate change, job insecurity and growing inequality must be tackled simultaneously, through greater grassroots economic democracy" (Earthworker n.d.). Earthworker's pilot project – Australia's first worker-owned green manufacturer – will produce solar-powered hot water systems and related components as part of a just transition in the wake of anticipated coal-fired power plant shutdowns in Victoria's Latrobe Valley.

Challenging the culture of consumption and adopting a "Simpler Way" as an act of resistance

Like the United States and other developed countries as well as the rising middle classes in developing societies, many Australians have adopted lock, stock, and barrel the capitalist culture of consumption. Furthermore, as compensation for its alienation in the workplace, much of the Australian working class has co-opted by the culture of consumption. According to Davis (2008: 138), "Australia now has more shopping centres per capita than any other Western nation – there are twice as many plazas and malls per head of population than the United States – and Australians spend more time there than in any other place outside home and work." Australian economists Clive Hamilton and Richard Denniss (2005: 7) in their discussion of *affluenza*, a manifestation of being unable to distinguish wants from needs, observe that along with the American pattern of households owning two automobiles, "other items that have become necessities in most Australian

homes in recent years are plasma-screen TVs, air conditioning, personal computers, second bathrooms, mobile phones and, increasingly, private health insurance
and private schooling for children." They also cite pet ownership as another manifestation of Australian affluenza:

> Sixty-four per cent of Australian households own one or more pets. We share
> our homes with an estimated 3.6 million dogs, 2.3 million cats, 7.5 million
> birds and 13.2 million fish. But growth in the number of pets accounts for
> only a small proportion of the growth in expenditure on pet food and pet-
> related products, which has been stimulated mainly by increased purchases
> of premium pet foods.
>
> *(Hamilton and Denniss 2005: 28)*

As neo-Keynesian economists, Hamilton and Denniss (2005: 7) blame neoliberalism as being responsible for promoting "higher consumption as the road to a better society" (Hamilton and Denniss 2005: 7). However, in reality, neoliberalism is
merely the latest phase of capitalism, and as Sklair (1991: 42) observed some time
ago, the culture-ideology of consumerism constitutes the "fuel that powers the
motor of global capitalism," an elite-generated ideology that promotes on-going
economic growth. Unlike Denniss (2017), who more recently proposes curing
affluenza by more closely regulating capitalism, I argue that a better solution in
the long-run, particularly given the gravity of the socio-ecological crisis, is to
transcend it with an eco-socialist vision for both Australia and the world at large.

While Australia generates one of the highest levels of greenhouse gas emissions
per capita in the world, high levels of consumerism in Australia also contribute to
greenhouse gases generated overseas, particularly in China, in terms of embedded
energy in the products being consumed, the emissions expended in manufacturing
them, and the emissions miles involved in transporting them (Holmes 2012: 398).

"More is better" permeates the housing sector too. In the mid-1950s, the average new house had about 115 square meters of space – that's half the size of houses
today. Between 1985 and 2000, the average floor area of new houses increased by
31 percent, from 170 square meters to 221 square meters while the size of apartments increased by 25 percent to an average of 139 square meters (Hamilton and
Denniss 2005: 20). Averages aside, there are vast class-based differences within the
Australian culture of consumption. A wealthy person with multiple residences,
such as a holiday house on the Great Ocean Road on the Victorian coast or the
Gold Coast in south-eastern Queensland, in addition to a spacious principal resi-
dence, leaves a much greater ecological footprint than someone who lives in a small
apartment or than a homeless person.

Ted Trainer (1998: 8) has long argued that developed societies such as Australia
exceed sustainable levels of production and consumption. Australians, he points
out, "can only live as affluently as we do because we are taking and using up most
of the scarce resources and preventing most of the world's people from acquiring
anything like a fair share" (Trainer 1998: 10). Many products sold in Australian

shops, such as electronic goods and clothing, have been manufactured in developing countries, particularly China, under working conditions that Australians would not tolerate at home.

Fortunately, a growing number of people in developed societies, including Australia, are opting to "downshift," reducing their consumption in myriad ways – though their numbers are not yet high enough to significantly contribute to climate change mitigation (Hamilton and Denniss 2005: 153–177). Like most climate change mitigation strategies, reducing consumption can further human well-being. Stilwell (2000: 56) maintains that while capitalism has delivered an impressive array of commodities, many people have come to recognize that "our principal sources of satisfaction – personal security, social recognition and interesting work – are often destroyed in the preoccupation with consumption." While much more than mere individual effort is required to mitigate climate change, for many of us, it starts with the individual, for climate justice requires that, one by one, hundreds of millions of the affluent people, including millions of Australians, scale back their consumption of material goods. While many working-class Australians turn to consumerism to compensate for alienation in the workplace and everyday social life, consumerism also correlates with class position: again, the wealthier the Australian, the greater the consumption, the greater the emissions.

Restraint in consumption is only reasonable and fair, partly in the case of developed societies. Trainer (1989: 199–201) calls for "appropriate development" for both "rich" and "poor" countries. For poor countries, the endgame is local self-sufficiency; the use of low, intermediate, and alternative technologies processing available resources; and a commitment to environmental sustainability (Trainer 1989: 196). For rich countries, an enormous reduction in consumerism is needed. Ideally, Australia and other wealthy countries would undergo de-growth and "work hard at reducing the amount of producing and consuming going on" (Trainer 1995: 108). Such a "conserver society," committed to great economic equality and with "much lower than present rates of per capita income and resource use," would help create a "world order that is peaceful, just, ecological sustainable and in which inequality and poverty have ceased to exist" (1991: 124). Trainer's (2010) notion of the "Simpler Way" is based on far simpler material living standards, high levels of self-sufficiency (within households, neighborhoods, towns, and nations), relatively little trade or transportation, small-scale cooperative economies in which most of the things we need are produced by local labor from local resources, much greater use of renewable sources of energy, and a commitment to human rights and social justice, particularly with regard to developing countries. His is an apt description of the principles of eco-socialism put into practice.

While many working-class Australians turn to consumerism as compensation for alienation in the workplace and everyday social life, consumerism correlates with class position: the wealthier the Australian, the greater the consumption, the greater the emissions.

Promoting sustainable settlement patterns, housing, transportation, and travel in a big country

Modern cities have evolved following the dictates of capital, with their need for manufacturing, financial, commercial, distribution, and communication centers, as well state bureaucracies. They are also one of the main drivers of climate change. Dawson boldly asserts:

> Not only are cities dependent upon nature, but they also structure our increasingly chaotic natural world. Climate change will unleash the greatest havoc in cities, but cities will also produce the most ferocious struggles against the inequalities of our urban age.
>
> *(Dawson 2017: 9)*

Virtually all cities today constitute "extreme cities" in that they are spaces of stark socio-economic inequality and ecological crisis, threatening the sustainability of urban existence in its present form. While some economists and urban planners argue that cities can be made "greener" and at least adapt to climate change, such benefits will inevitably accrue to the affluent and powerful as opposed to the teeming massing of poor, particularly in the case of developing countries, living in them. Dawson argues that the future for urbanites will be wrenching and will likely require a massive retreat from cities and radical revamping of the present geography and global cities. He argues that reversing the trend toward extreme cities, which are largely a by-product of global capitalism, will entail reversing the deteriorating social and environmental conditions of rural areas, so that people can choose to remain in the countryside without being subjected to socio-economic deprivation on numerous fronts.

In response largely to a looming climate crisis, a new urbanism emerging around the world seeks to make cities more environmentally sustainable and thereby also more livable. However, much of the new urbanism is framed within the capitalist parameters of economic growth and accumulation. As Frankel (2018: 77) observes, "for nations, and the world as a whole, *new environmentally sustainable cities presuppose new alternative political economies.*" Thus, what is needed is a new radical urbanism which addresses not only issues of sustainability but also social justice and inclusion that challenges the gentrification all too often associated with the new urbanism. In his next book, Frankel correctly argues:

> [T]he new urban imaginary can only be achieved by political reform or even radical action at state level that requires instituting legislative changes to convert private property and publicly owned land into green commons. This would require reversing the privatisation of public roads, utilities and other assets acquired by corporations in recent decades. To make a large city sustainable would involve challenging the sacred notion of the ownership of private property which is the foundation of capitalism.
>
> *(Frankel 2020: 368–369)*

Housing

Homeownership is a cornerstone of Australian society. Australian government policies presently encourage people to remain in their houses as they age, after their children have flown the coop. Seamer observes:

> While this is good for many, it has the clear disadvantage of underutilising large suburban large suburban homes inhabited by one or two elderly residents rather than groups of people such as families.
>
> *(Seamer 2019: 134)*

Furthermore, many of these senior citizens find themselves socially isolated.

Australian cities developed as regions of urban and suburban sprawl, committed to the notion that each household have a quarter-acre block and one or more cars to transport its occupants through the resulting sprawl. While homeownership has historically been widespread in Australia, it has never been universal. The median house price in Melbourne, for example, rose over 70 percent between 2012 and 2018, and the median apartment unit price rose 30 percent over the same period (Mares 2018: 17). Mares (2018) reports that 26 years of housing growth and skyrocketing property wealth "have been accompanied by higher levels of homelessness, more housing insecurity, declining rates of home ownership and greater housing stress in Australia." Australia reportedly has 1.9 million property investors, roughly 15 percent of all taxpayers who profit from existing properties, thereby inflating property prices without adding to the stock of housing (Megalogenis 2016: 51). Modest livable housing should be a right for all, along with education and health care. In order to achieve more equitable housing, state governments need to construct public medium-height housing for low-income people, not atomistic 20-story housing developments of the past.

Australia once had a more extensive public housing sector, which "used to serve a much broader segment of the populace, including working families" (Gleeson 2010: 36). This sector needs to be rejuvenated and expanded by, among other initiatives, providing affordable housing for low-income people in the increasingly gentrified inner suburbs of Australia's cities. Public-funded co-housing arrangements in which housing would be dictated by needs could potentially lead to more equitable living arrangements and a sense of community for people of varying ages.

Since World War II, Australians have followed a North American pattern in terms of the size of their houses, particularly in outer suburb areas where large numbers of McMansions have appeared. Conversely, many affluent residents in inner-city suburbs live in huge houses. Phibbs and Gurran report:

> Australian dwellings are able to accommodate an increase in household size because of the large size of Australian houses – we now have the second largest houses in the developed world. Between 1960 and 2010 the size of new homes in Australia more than doubled, although since then they have

reduced by about 10 percent as apartments have become much more popular. In 1970, about 13 per cent of dwellings had four bedrooms compared to 30 per cent in the 2016 Census.

(Phibbs and Gurran 2018: 209)

Unfortunately, due to the increasing cost of housing, in terms of both ownership and rental, a growing number of Australians have become homeless. Indigenous people are much likely to be homeless with "comparative rates of homelessness over ten times that of non-Indigenous Australians" (Habibis and Walter 2015: 232). From an eco-socialist perspective, everyone should have access to decent housing. When families diminish in size as children grow up and leave home, perhaps eventually to start their own families, older couples can move into smaller quarters. Holmes (2021: 10) calls for a ban on foreign ownership of housing in Australia and maintains that the "state should build quality public housing on a massive scale to end homelessness, house everybody who needs it and end capitalist property speculation and rent gouging."

In recent years, more and more Australians residing in large cities, particularly Sydney and Melbourne, live in high-rise apartment buildings, some in the order of 50 stories high. With rapid-speed lifts, air-conditioning, and other amenities, such structures tend to be highly energy-intensive and encapsulate a lot of embedded energy in terms of steel and concrete and contribute to atomistic living (Graham 2016). Skyscrapers, whether they are high-rise apartments or office buildings, are "profligate users of fuels for light, power, and services" (Day 2020: 119). While precise figures are not readily available for Australia, the

UK Green Building Council estimates that the construction industry generates about 22 percent of UK carbon emissions, uses 40 per cent of the drinking water, contributes 50 per cent of climate change emissions and over half our landfill waste, and accounts for 39 per cent of global energy use.

(Day 2020: 122)

One wonders what would happen to residents on the upper floors of these apartment buildings in a power black-out and/or heatwave. A far more livable and environmentally sustainable and socially amicable form of urban living would be medium-density, low-rise housing which "works very well for many parts of European and some older inner-city Australian suburbs" (Mackay 2018: 49). Conversely, Australian suburbs, in many ways resembling many North American suburbs, consist of quarter-acre blocks "may potentially increase the risk of social isolation" (Mackay 2018: 50).

Given that the outer suburbs exist, some scholars have come to think about how to make them more livable and environmentally sustainable. Gleeson (2010: 190 has suggested that a low-density suburb "might have a relatively modest ecological footprint" if residents opted to restrain their consumption of goods and services, use their automobiles when necessary, grow some of their food, collect and treat water, and shop and socialize locally. In keeping with his call for a Simpler Way, Trainer maintains:

> [H]ousing design, especially location and functions, must maximize the community's material, social and spiritual "productivity" in ways that all the world's people could share. The need for large-scale degrowth obviously implies a shift to housing that is as small, cheap and humble as is reasonably possible. This would strike most people in consumer society as totally unacceptable. Thus the task for us is to show that sustainable housing for all the world's people does not have to involve deprivation or hardship, indeed that it can be delightful. For a start, it makes housing for all, whereas currently housing is a commodity enabling wealth-seeking speculators to get richer without having to work while, even in the richest countries, large numbers of people cannot afford a house.
>
> *(Trainer 2019: 124)*

A small number of people in Australia have been exploring co-housing, eco-communities, and eco-villages as more socially cohesive and environmentally sustainable housing arrangements, although unfortunately some of these pre-figurative experiments are largely the preservative of relatively affluent people residing in inner-city suburbs (Nelson 2018; Cooper and Baer 2019).

Besides some downsizing for the rich and more equitable housing for all, tending community gardens and family unit gardens has a role in addressing environmental problems and mitigating climate change. Alongside cities in other developed and developing countries, cities such as Brisbane and Melbourne are fostering community gardens but unfortunately doing little in terms of providing affordable housing for low-income people. Some ground-up ideas have started to permeate urban planning in Australia. Rob Adams (2009), Melbourne's director of City Planning, proposes situating medium-density housing development in specified "urban corridors" requiring all developers to make 25 percent of all new housing in any residential development affordable, and streets everywhere are to be modified by government bodies to favor rapid public transportation, bicycles, and pedestrians over automobiles.

Transportation

Australia, like the United States and Canada, is car-oriented, not to say car crazy. Australian cities developed as regions of urban and suburban sprawl, committed to the notion that each household have a quarter-acre block and one or more cars to negotiate the resulting sprawl. As a result, commuting and moving around in the larger cities is increasingly difficult, even where public transport infrastructure is elaborate, because much more money has been poured into road-building than into developing and upgrading public transportation, further encouraging car use. Car-dependency is especially acute in the outer suburbs, so an immediate priority for the federal government and state governments should be to improve public transport services there, matching the quality found in inner and middle suburbs. Urban sprawl has made commuting and moving around in the larger cities increasingly difficult, even where elaborate public transport infrastructure exists, because much more money has been poured into road-building than developing and upgrading public transport, further encouraging car use. Car-dependency is especially acute in outer suburban areas, so any immediate priority for governments should be to improve public transport services to these areas to match the quality found in inner and middle suburban areas (Matthews 2010).

The Rapid and Affordable Transport Alliance, which consists of 17 environmental, health, and union bodies (including the Australian Conservation Foundation and the Australian Council of Trade Unions), published a report recommending that the federal government devote two-thirds of transport funding and only one-third to roads during the following decades (Matthews 2010: 12). The report noted that between 2004 and 2009 the federal government had spent $1.2 billion on rail but $14 billion on roads. Within and between urban centers, energy-efficient and environmentally sustainable forms of transportation are needed for people and freight. Australian transportation expert Peter Newman (2009: 108) observes: "The biggest challenge in an age of radical resource-efficiency requirements will be a way to build fast rail systems for the scattered car-dependent cities." In a similar vein, Gleeson argues:

> The Commonwealth will need to drive a complete renovation of our transport policies and support the metropolitan commissions in this work. Australia must discard transport policies that massively favour private motorised travel over public transport and non-motorised forms of transport such as walking and cycling.
>
> *(Gleeson 2010: 134)*

Canberra, designed by American Walter Burley Griffin, is emblematic of an unsustainable vicious cycle in transport and urban planning. A national capital city with no rail link to any state capital except for Sydney (and a very slow one) is clearly the world's worst practice, although a short light rail line was open several years ago between Civic and Gungahlin, an outer suburb. While Melbourne has one of the world's widest-flung suburban train systems and the world's most extensive tram system, infrastructural and technical difficulties keep it from realizing its full potential in mitigating climate change and serving citizens. Buses and trams are slowed greatly by congestion caused by automobiles. In Melbourne and other Australian cities, complete socialization of the public transportation system, coupled with an increase in train carriages, trams, and even buses, would help to reduce fares and spur additional ridership.

Nicholas Low, a Professor of Urban and Environmental Planning at the University of Melbourne, proposes two principles of "sustainable transport" that are congruent with an eco-socialist perspective vis-a-vis both social justice and environmental sustainability, albeit with some qualifications. His first principle:

> For an urban transport system to be sustainable it must be one in which the carbon emissions from its operation, and embodied in new infrastructure construction, are reduced to a level compatible with a global temperature rise from pre-industrial levels of no more than 2 degrees Celsius.
>
> *(Low 2013: 70–71)*

Although some eminent climate scientists, including James Hansen, have recently come to believe that a lower limit of 1.5°C limit above pre-industrial levels is needed to achieve a safe climate, Low's general definition merits a close look. Low's second principle of sustainable transport states:

> For a sustainable urban transport system also to be fair it must be one in which per capita carbon emissions are approximately the same in all cities worldwide.
>
> *(Low 2013: 70)*

Automobiles that remain with us need to be more energy-efficient and as small as possible. Large four-wheel drive vehicles should be banned from urban areas, where their use has become not just an ecological problem but an increasing hazard for other road users and pedestrians. Air travel for business purposes should be restricted via intelligent use of technological alternatives to face-to-face meetings. More sustainable forms of holidaying would entail more trips closer to home, ideally by means other than a private motor vehicle, and fewer journeys to distant lands.

Public transport utilization in Australia tends to be uneven. Stock, Steffen, Bourne, and Brailsford report:

> Many public transport systems in our major cities, particularly light rail and train services, are bursting at the seams. Many public transport services are already at or beyond capacity during peak periods, leading to crowding, delays and worsening quality of service.
>
> *(Stock et al. 2018: 22)*

Conversely, they report:

> [M]any bus services are characterised by low use and low levels of satisfaction. Issues affecting bus services and levels of use include indirect routes, infrequent services, limited hours of operation and poor coordination with trains and light rail.
>
> *(Stock et al. 2018: 22)*

In addition to investment in new high-quality public transport systems, what is needed is public operation and centralization of public transportation. For example, in Melbourne, private companies operate public transportation, namely the Metro suburban trains, Yarra Trams, and a slew of small companies operating buses whose schedules are poorly coordinated with train and tram schedules. Fortunately, almost all capital cities in Australia are planning, building, or extending new light rail or train services. The Gold Coast Light Rail system opened in 2014, connecting 16 stations along 13 kilometers of track and reduced motor vehicle traffic on the Gold Coast Highway (Stock et al. 2018: 43).

Due to the size of the country with a population of some 25.7 million people and with an array of mediocre railway systems between major cities, Australians have a strong propensity to both fly within the country and overseas. Australia's transportation is part and parcel of an aviation industry that had been prior to the recent COVID-19 pandemic on a long growth spurt since after World War II both in terms of transporting people and cargo. Up until the outbreak of the pandemic, Ede reports:

> Today there are between 9,000 and 10,000 commercial aircraft in the sky at any moment, and between 600,000 and 1 million passengers in flight. Between 1974 and 2015, global passenger traffic rose from 400 million passengers annually to almost billion passengers. Impressive as the passenger statistics are, they are surpassed by the growth in air cargo transport. In 1974, 16,955 million tons of cargo were transported by air. In 2015, 195,162 million tons were transported and the delivery company FedEx was operating the fifth largest fleet of commercial aircraft in the world.
>
> *(Ede 2019: 272)*

As Green and Minchin (2010: 107) observe, flying constitutes the "main way for Australians who need to get out of the country for work or who want to see the world." The Brisbane, Sydney, Perth, and Melbourne airports have expanded their facilities and runways in recent years, and the federal and state governments are providing financial support for the creation of a second major Sydney airport (Lohmann and Trischler 2017). Australia has the potential for improving upon its existing railway system by developing renewable energy-power high-speed rail, expanding existing state railway systems, and nationalizing the Great Southern Railway system. Queensland Rail, New South Wales Train, and Victoria Line operate relatively extensive state rail systems which include connecting bus lines that could be expanded. For the past several decades, many regional communities gave up train lines as well as post offices (Denniss 2019: 95).

The Western Australian railway system could be expanded beyond its present lines connecting Perth with Kalgoorlie to the east and Perth with Bunbury to the south. The Great Southern Railway consists of three lines: The Indian Pacific, which connects Sydney to Broken Hill to Adelaide to Kalgoorlie to Perth; the Ghan, which connects Adelaide with Alice Springs and Darwin to the north; and the Overland, which connects Melbourne and Adelaide, a train which I have taken several times. The Overland unfortunately runs in each direction twice a week. The Indian Pacific and the Ghan also have limited runs and by and large serve an affluent tourist clientele. Unfortunately, since World War II, "road haulage has replaced rail transport as the main interstate carrier" (Catley 1996: 119). The nationalization of the Australian transport system has the potential of creating a more environmentally sustainable way to transport both passengers and freight across a huge country. A consortium consisting of Beyond Zero Emissions, the Melbourne Energy Institute, and the German Aerospace Centre (2014) proposed a solar-powered high-speed passenger rail service connecting Brisbane, Sydney, Canberra, and Melbourne. Highlights of the proposed system include the following points:

- Containment of 45 percent of Australian regional travel within the network corridor;
- 21 stations connecting 13 regional centers and six major cities;
- Journey times of less than 3 hours from Sydney CBD to Melbourne or Brisbane CBD;
- 60 percent of the Australian population situated within 50 kilometers of a station on the network;
- Three million fewer domestic passengers at Sydney airport in 2030 than current levels;
- The dominant transport mode for a journey between 350 and 1,300 kilometers in distance (Beyond Zero Emissions, the Melbourne Energy Institute and the German Aerospace Centre 2014: viii–ix).

Australia is a country with heaps of tourists, both domestic and international, who tend to fly to distant places in a far-flung country on airplanes but may also drive in motor vehicles. Prior to the COVID-19 pandemic, many affluent Australians engaged in lots of overseas travel, particularly to Europe and North America, but also to Indonesia, India, Southeast Asia, the Middle East, and China. Much of this travel, generally in the form of tourism, constitutes a form of experiential consumerism. A Simpler Way would entail a shift to local and slow travel. One often hears the argument that overseas travel contributes to cosmopolitanism, but in reality Australians can become quite cosmopolitan by learning more about and interacting more with the many ethnic and nationality groups within their own society beyond and above simply dining in restaurants offering the cuisines of the group. Furthermore, cosmopolitanism or learning about diverse cultures can be achieved through reading and films, the latter constituting a form of cyber travel.

Sustainable agriculture and forestry

Living on a "sunburnt land," in contrast to Indigenous Australians, who occupied a vast continent for some 50,000–60,000 or even more years, the vast majority of

Australians as settler-colonists and immigrants have always faced incredible environmental problems which are sure to exacerbate due to climate change over the course of the 21st century. Among these environmental problems, one can include water shortages, deserts, soils with low nutrient levels, low and unpredictable rainfall, depletion of aquifers due to irrigation, salinization, depletion of old-growth forests, and marine overfishing (Diamond 2005: 380–406). Can Australians feed and shelter themselves while trying to mitigate climate change? In 2019 Australia was one of the world's largest consumers of beef, averaging 25 kilograms per capita (Meat & Livestock Australia 2020: 7). In 2019, Australia's lamb consumption was 11 kilograms per capita.

With emissions from Australian agricultural practices especially significant, an important form of climate change mitigation would be a decided shift in food production from heavy reliance on meat, particularly livestock in the form of cattle and sheep. Carrington reports:

> [N]ew research shows that without meat and dairy consumption, global farmland use could be reduced by more than 75% – an area equivalent to the US, China, European Union and Australia combined – and still feed the world. Loss of wild areas to agriculture is the leading cause of the current mass extinction of wildlife. The new analysis shows that while meat and dairy provide just 18% of calories and 37% of protein, it uses the majority – 83% – of farmland and produces 60% of agriculture's greenhouse gas emissions.
>
> *(Carrington)*

According to Bert Metz (Metz 2010: 242), "changing to a vegetarian diet can avoid N_2O emissions from grasslands, CH_4 emissions from livestock and manure, and CO_2 emissions from fossil fuel use, and free land for other purposes." Due largely to industrial agriculture, which relies heavily on gasoline-powered farm machinery, irrigation, chemical fertilizers, and pesticides, as Lowe (2016: 80) observes, "Australia currently produces about enough food for 50 million people, exporting more than half of that production." But what kind of food is being produced? Given heavy emissions from Australian agricultural practices, an important climate change mitigation strategy would be to shift production away from a heavy reliance on meat, particularly livestock, and toward mixed crops and plants. Fortunately, Australia has a "multitude of local organic and permaculture groups and biological and regenerative farming networks" which constitute an important base for developing more sustainable agriculture (Broughton and Garcia 2017: 48). This is a tradition that needs to be greatly expanded. Gleeson (2010: 119) advocates local food production, noting that Australia's "major cities once sourced much of their food in their immediate hinterlands."

In comparison to the United States, where cattle production occurs in feedlots requiring massive amounts of grain, most Australian cattle graze on grasslands, with only about 2 percent of Australia's cattle production occurring on feedlots (Broughton and Garcia 2017: 59). However, about 10 percent of Australia's greenhouse gas emissions come from livestock production, mostly due to methane produced by natural digestion on the part of cattle and sheep. Substitutes for beef and lamb in Australia would be kangaroo as a lean and relatively healthy red-meat alternative, and rabbit as a white-meat alternative, both of which could be obtained through periodic culling and would decrease emissions from food production and health problems for humans.

Where animal production continues, methane from animal digestion could be reduced through practices such as vaccinations and chemical inhibitors; and changes in tillage practice and vegetation cover. It may also be possible to reduce NO_2 emissions from fertilizers through fertilizer management, soil and water management, and fertilizer additives (Rickards and Tucker 2009: 93). Last but not least, bioregionalism in terms of food distribution and consumption could potentially prevent millions of "food miles," creating emissions by the unnecessarily long-range distribution and long-term storage of food products.

Forests need to be part of eco-socialism's energy calculus too. Australia's forests – particularly old-growth forests – must be preserved to absorb some of the carbon that the capitalist system is pumping out. Since European settlement, forests on the east coast and in Tasmania have been decimated due to clearing for farm land (Steffen et al. 2009: 32). In his chapter on " 'mining' Australia," Jared Diamond discusses the "double irony" of the Australian forestry industry:

> On the one hand, Australia, one of the First World countries with the least forest, is still logging those shrinking forests to export their products to Japan, the First World country with the highest percentage of its land under forest (74%) and with that percentage still growing. Second, Australia's forest products trade in effect consists of exporting raw material at a low price, to be converted in another country into finished material at a high price and with high added value, and then importing finished products.
>
> *(Diamond 2005: 404–405)*

Excessive tree-clearing in Queensland and elsewhere has proceeded despite growing awareness of the need to reduce carbon pollution. Forest plantations, as Lester Brown (2009: 195) argues, can reduce pressures on the earth's forests, including Australia's, but only as long they do not replace old-growth forests.

Population and immigration policies favoring the most-needy people in developing countries

Demographers have often observed that the greatest rate of population growth tends to occur among the poor, especially in developing countries, where people regard children as economic assets because they work the land or sell products or services in the cities. Distributing economic resources more equitably would help slow population growth, particularly if household economic improvements are accompanied by policy changes giving women more control over fertility decisions. At the international level, reversing the trend toward increasing economic inequality within countries and between countries, and improving the rights and status of women, clearly constitute the best means to reduce population growth.

Population growth constitutes a ticklish issue for many Australian environmentalists, such as Tim Flannery, who has argued that a sustainable population for Australia would be somewhere in the order of 6–12 million (Paul 2016: 89). In Australia, the population issue is inextricably linked with problems of affluence rather than poverty. Most Australians are consuming many times more resources than do the vast majority of people in developing countries. When in 2010 the government announced its projection that the Australian population could increase from 22 million people at the time to 35 million by 2050, some environmentalists shot back that such a population would not be sustainable in a country whose resources are being strained by both urban growth and drought. That's undoubtedly true if today's overall consumption level continues or increases. O'Connor and Lines (2008) imply that population reduction is desirable because it would permit affluence-as-usual. They admit that, "if the world's resources were ever shared equally, no one would have a first world life-style," but if the world's population could somehow be reduced to 2 billion "then there would be enough resources for everyone to live a European material standard of living (standard of consumption) which is about half of that of the USA" (O'Connor and Lines 2008: 28). However, an alternative Australia based upon eco-socialism and committed to a simpler way of living and redesigned for a sustainable living could support a larger population.

Population growth in Australia, as is the case for most other developed countries, is driven more by immigration rather than by natural increase. Immigration has played a pivotal role in the development of Australian capitalism. In the post-World War II era, immigration to Australia was "primarily focused on supplying labour for the industrial and manufacturing centres of Australian capital," but in more years it has become more diversified, focusing on bringing into the country more white-collar workers, middle-class professionals, along with "increased number of small and large business-owning migrants bringing with them significant capital and assets" (Humphreys 2019: 30). In addition to generating revenue for Australian universities, the international student program seeks to entice the higher-performing students into the Australian labor force (Humphreys 2019: 33).

An eco-socialist perspective on population growth and concomitant environmental strain could counteract Australia's widespread xenophobia, including Islamophobia, related to refugees and other immigrants. A shift in thinking is critical since immigration accounts for more of Australia's growth than natural population-replacement increase does. Although debates about ecological footprints inevitably include anti-immigration voices, some climate movement spokespersons have sought to counteract widespread hostility in Australia toward refugees by arguing for increasing the humanitarian component and decreasing the business component of Australian immigration policy. Lowe observes:

> A goal of making Australia a truly lucky country would require us to be prepared to share our luck with the poorest people in the world. Floods of

refugees from Afghanistan and Iraq are arguably direct consequences of our involvement in ill-judged military operations in these countries. Instability and poverty will continue to motivate people to try to get away from where they are, in search of better lives and for them and their children. The problem will certainly be compounded in coming decades by "climate change refugees", forced to leave Pacific islands or low-lying areas on the Asian mainland and the Indian sub-continent as rising sea levels make their living conditions impossible.

(Lowe 2016: 82)

The Greens advocate that the admittance level of refugees be raised to 50,000 per annum, but this figure could be easily increased if the political will were present. As McIntosh (2020: 162) asserts, "for all the uncertainty around climate change as a current driver of refugee flows, people's need to find new homes in habitable parts of the world is likely an emergent issue." Thus, in the name of climate justice, Australia, along with New Zealand, needs to adopt a more welcoming stance toward peoples displaced from low-lying islands in the South Pacific, such as Kiribati and Tuvalu. A climate justice perspective demands that Australia offer climate asylum to people of the world considerably less complicit than average Australians in causing climate change. Immigration policy should focus on permanent migrants rather than temporary migrants. Jobs for citizens should be prioritized, thus preventing employers from hiring cheap labor.

Partial de-linking from the capitalist world system

Achieving an eco-socialist Australia would have to be part and parcel of a larger global effort in which anti-systemic movements, including a vibrant climate justice movement, would push for a more socially just and environmentally sustainable world system. Australia appears on its way to becoming an "80/20 society," one in which 20 percent of the population lives on the margins and the other 80 percent manages, varying from very well to barely. Unfortunately, as in the United States, the seductive culture of consumption – with its hallmarks of large residences and private motor vehicles, and trips overseas – reigns supreme in Australia. Breaking free of this headlock requires an alternative vision.

Writ large, this process of letting go is what Samin Amin (1990) calls *de-linking*. Delinking, the gradual move toward socialism by disengaging with global capitalism, theoretically allows a post-revolutionary society to achieve national development outside of the imperatives of the capitalist world system. Unfortunately, delinking, as Halliday (1999: 282) observes, "has throughout history been associated with dictatorship, frenzied nationalism, and the cult not just of the nation and its supposedly legitimating past, but often also of individual leader," as was the case with Stalin in the Soviet Union, Mao in China with the Great Leap Forward (1958–1961) and Great Proletarian Revolution (1965–1971), and Kim Il Sung in North Korea with the policy of *Juche* or "self-reliance." Unfortunately, under Stalin, the Soviet

Union abandoned the path of socialism for a bureaucratic collectivist dictatorship. However, China – unlike almost equally populous India at the time – was able to undergo rapid economic development by de-linking during the Maoist era and much more effectively eradicate much abject poverty. More recently, Amin (2013: 143) reintroduced the notion, making *de-linking* synonymous with *de*-globalization, and putting it forth as a "strategic reversal in the face of both internal and external forces in response to the unavoidable requirements of self-determined development" and as a mechanism for "reconstruction of a globalization based on negotiation, rather than submission to the exclusive interests of imperialist monopolies" and contributing to the "reduction of international inequalities."

De-linking is not for the faint of heart. Amin asserts that it can be "initiated only in the context of states/nations with advanced radical social and political struggles, committed to a process of socialization of the management of their economy" (Amin 2013: 143). Aside from the open question of whether societies such as the Soviet Union and China were ever socialist or something else, COMECON (Council for Mutual Economic Assistance) among the Soviet-bloc countries as a semi-autarkic structure no longer exists and China has become so integrated into the capitalist world system that some neo-Marxian scholars would argue that it is a state capitalist society.

Given the wrong turns and surprises in socialism's history, perhaps gradual de-linking should be considered as a transitional strategy for breaking the capitalist world system's stranglehold on energy and other natural resources in countries which had already tried to create socialism for the 21st century. As Frankel (1992: 281) observed some time ago, a "semi-self-sufficient Australia would need to change its production and consumption patterns radically in order to achieve genuine self-determination." More recently, he has argued that "[s]emi-autarky or substantial rather than full self-sufficiency is possible with controlled levels of imports and balancing increased local production of needs." Only time will tell whether blocs of partially delinked countries in the developed world could form, including one in southern Europe consisting of Greece, Italy, and Spain, which have increasingly found themselves marginalized within the European Union.

While it may appear far-fetched right now, another delinked bloc among the developed countries might include Australia and New Zealand, along with Papua New Guinea, Timor Leste, and various South Pacific island-states (many of which have abundant mineral resources and agricultural land). Currently, Australia and New Zealand constitute either minor core countries or relatively affluent semi-peripheral countries, but Papua New Guinea, Timor Leste, and the South Pacific island states are poor peripheral countries. That needs to change.

In the energy realm, given that although Australia relies heavily on coal for domestic energy, it could make a drastic shift to reliance on renewable energy sources, particularly solar, wind, and geothermal, all of which exist in abundance in the country, particularly if a new left party were to come to power ready to nationalize the energy sector and orchestra such a shift. While the manufacturing sector has slumped in Australia, it has a highly educated and skilled labor force that could manufacture a wide variety of products for domestic consumption – ranging from cars to clothing and electronic appliances.

As a result of corporate globalization, Australia has come to rely for many years over its consumer products from overseas markets, a scenario that was not the case nearly as much 50–60 years ago. In 1937, 47.2 percent of the Australian work-force was employed in factories. Consequently, during World War II, Australia was highly self-sufficient in food production and manufactured products, including military equipment for the war effort (North 2016: 328). Lowe reports:

> In 1964 Australia produced radios and television sets, electronic equipment for communications and navigation cars and larger transport vehicles, boats and light aircraft. Over the last 50, we have systematically run down our manufacturing capacity. The last large-scale vehicle building plants will close in the next few years [The scenario which has transpired]. Appliances and electronic equipment in our stores mostly come from overseas. We even import such basic products as clothes and shoes. Our escalating import bill for advanced manufactures has been covered by exporting ever-increasing quantities of unprocessed raw materials, principally iron ore and coal. The only change is that the exports now are more likely to go to China, Japan or Korea than the UK or other European countries; equally, the imports of labour-intensive goods like textiles, clothing and footwear are now predomi-nantly from China, while the imports of knowledge-intensive products like cars or computers are mostly from Japan or Korea.
>
> *(Lowe 2016: 65–66)*

The complex system of exports-imports whether by ship or airplanes is highly dependent on oil, which results in greenhouse gas emissions and is yet another driver of anthropogenic climate change. As Pettifor (2019: 74) observes, "While trade will always be a characteristic human interaction, on the scale undertaken today, and under the terms prevailing, it is not sustainable. Nor is it socially or eco-logical just." Shipping alone, upon which Australia is highly dependent in terms of exporting mineral resources and agricultural products and importing a wide array of products, ranging from machinery and motor vehicles to electronic equipment, clothing, and boutique food items, accounts about 2 percent of global carbon diox-ide emissions along with ocean pollution (Frankel 2018: 107).

Thus, there is a need for Australia, along with other countries, to shift as much as possible to localize their economies, something that flies in the face of global capitalism. The same holds true for African countries, be it South Africa, Nigeria, Ethiopia, Kenya, etc. African-born Pettifor aptly argues:

> The continent itself is one of the richest in terms of natural resources. That is what make it such a target for exploitation by richer countries. Africa is not poor, it is impoverished. In other words, it has been made poor, mainly by

free trade, liberalisation and privatisation, all of which exacerbate the massive losses, extraction and slavery associated with colonial exploitation.

(Pettifor 2019: 77)

Redirecting manufacturing this way would also help the country reduce imports, which presently stand at 88 percent of manufactured products consumed in Australia (North 2016: 57). As a relatively resource-rich country with a small population, Australia should rely more on its own domestic resources. Export-oriented countries are highly vulnerable to downturns in the capitalist world economy, particularly to crises or quick shifts among its major economic players, including the United States and China. In any case, blocs of de-linked countries moving toward socialism could form a super-bloc that would stand in opposition to the capitalist world system, which is integrated and supported by such structures as the World Bank, IMF, WTO, the World Economic Forum, G7, and G20.

Reducing Australian reliance on US military hegemony would include shutting US-operated intelligence facilities on Australian soil, such as the joint facility at Pine Gap in the Northern Territory. By allying itself with the United States in wars in Afghanistan and Iraq, Australia has contributed to the refugee crisis, including its own, meanwhile maintaining a very draconian refugee detention process on Australian soil and in other locations – Christmas Island in the Indian Ocean, Manus Island in Papua New Guinea, and Vanuatu in the South Pacific. The current Australian-US alliance threatens regional stability" and compromises Australia's strategic independence, as is evidenced by the fact that Australia have followed into every major conflict beginning with the Korean War. Even former Coalition Prime Minister Malcolm Fraser eventually concluded that the Australian-US defense alliance needed to be scrapped while other critics have suggested that Australia maintain neutrality in the same vein that countries such as Switzerland and Sweden have done. Such a position seems prudent given the growing geopolitical tensions between China and the United States, particularly during the Trump presidency of 2017–2021. Indeed, Richard Di Natale while he served as the Greens leader argued that the Australia-US military alliance constituted a security threat to Australia and called for the abandoning the alliance.

Some Australian politicians and political pundits in recent years have come to view China as a possible military threat, even though it constitutes Australia's leading trading partner. However, in contrast to the United States, the UK, and even Australia, China has not been in the habit of invading other countries. While China indeed has increased its military power, much of this is a defensive stance as it has come to view itself as facing a belligerent bloc consisting of the United States, Japan, and Australia. Toohey argues:

> Australia is one of the most secure countries on earth. It doesn't share land borders with any country, let alone any country with a history of ethnic, religious or other hatred stretching back centuries. It has not been invaded since 1788 [when Captain Cook and his crew landed on Botany Bay in what is

present-day Sydney]. . . . It would be good if the US became less belligerent and China's didn't behave like the kind of hegemonic power it once decried. But we have to deal with the world as it is – a world with almost no threat to Australia but a genuine risk of a military conflict between the US and China.

(Toohey 2019: 328)

Over the past several years, an anti-China campaign has emerged in Australia which for the most part is overstated (Hamilton 2018; Hamilton and Ohlberg 2020). Although China constitutes a rising superpower which began to displace the Soviet Union three decades ago, its influence in the Australian political economy does not match that of the United States and even still the UK. In terms of foreign investment in Australia in 2019, US investments came to $993.7 billion, UK investments came to $686.1 billion, and trailing Belgium, Japan, Hong Kong (SAR of China), Singapore, the Netherlands, Luxembourg, came China proper in the ninth position at $64.4 billion (see statistics in Holmes 2021: 10).

Ultimately, delinking would be only a short-term strategy that would tide Australia and other prospective bloc members over until a groundswell of socialist-oriented regions and countries achieved a socialist world system powerful and united enough to edge out today's capitalist world system. Under a global eco-socialist system, as Sarkar observes,

> [T]he minimal long-distance trade between its various regions that would be necessary and possible within the limits of sustainability would be just, fair exchange. Perhaps it would then be conceivable to form a world economic council that would plan such trade. One of the objectives of such planning would be to distribute the world's resources equitably, transferring resources from richly endowed regions . . . to poor endowed regions.
>
> *(Sarkar 1999: 221–222)*

Conclusion

The radical, system-challenging steps that I have delineated constitute loose guidelines for shifting Australia toward an eco-socialist vision, one in dialogue with eco-socialist dialogues in other countries, both in the Global North and the Global South. I do not purport that my suggested guidelines are comprehensive because undoubtedly others could be added to the list. A government that has been democratically elected and truly expresses the will of the majority of the people would have the right to impose regulations that would promote social parity and environmentally sustainability, including measures to promote a safe climate, but that in all likelihood would be opposed by corporations and wealthy individuals. Trainer (2010: 299) advocates a "thoroughly peaceful revolution," which in theory is preferable, but may not always be possible given the political situations in which the capitalist class and its political allies seek a violent counterrevolution.

A broader question is whether societies, including Australia and the world system based on eco-socialism, would allow room for markets. In my vision of eco-socialism, new left parties that rise to power in democratic elections would in time nationalize or socialize the means of production, particularly large multinational and national corporations that today have a powerful influence, both directly and indirectly on governments and mainstream political parties. I can see room for small businesses, such as restaurants, cafes, bookshops, clothing stores, cinemas, experimental theaters, and so forth, preferably worker-owned and worker-managed, but possibly also owned by entrepreneurs or families. In an eco-socialist world, markets would not completely disappear but would be much diminished to their strong presence in capitalist societies, such as Australia. In Schweickart's (2016: 6) conception of a socialist-oriented *economic democracy*, small businesses would "provide jobs for large number of people, and goods and services to even more," as they presently do. Unfortunately, most small businesses today pay their employees low wages and provide them with minimal social benefits. Such practices would be eradicated under an eco-socialist model. Ultimately, eco-socialist-oriented societies, including in Australia, and the world system would have to create legislation that would have to restrict the size of businesses, lest the capitalist tendency for bigger businesses and concentration to take over the smaller businesses gain ascendancy.

7

AUSTRALIA AND ECO-SOCIALISM IN A GLOBAL CONTEXT

Is my vision of an eco-socialist agenda, partially informed by eco-anarchism, eco-feminist, de-growth, and Indigenous visions, for Australia wishful thinking? Perhaps the vision won't unfold by itself or soon. But if Australia, like other wealthy countries, keeps up its wholehearted embrace of the capitalist growth paradigm that's behind deepening social divides and a climate crisis worsening despite the deliberations at 26 UN conferences and despite emissions-trading schemes around the world, the alternative may be a dystopia. Lacking strategies that seek to transcend global capitalism, both Australia and the world as a whole face the dystopian view expressed by Rod Quantock in my dialogic conversation with him. While humanity still finds itself in the midst of a global COVID-19 pandemic, hanging on the hope that a variety of vaccines will allow it to return to business-as-usual or at least the way things existed prior to the outbreak of the pandemic, climate change lingers in the background around the world, including in Australia. Along with the climate crisis, the coronavirus (COVID-19) pandemic illustrates the deep-seated contradictions of capitalism. While the COVID-19 crisis ravages the health of human populations, resulting in diminished health and even death, in some countries more than others, multi-national corporations and governments around the world, including the United States, Europe, China, Australia, and Brazil, are biting at the bit to ramp up the capitalist world system in order to return to its previous treadmill of production and consumption, still heavily dependent on fossil fuels. The sports industry is anxious to restart its playing seasons, even belatedly, and at least have its teams play to empty stadiums with revenue generated from television viewership as a prelude to returning to business as usual as teams and fans scurrying, often by airplane, across their home countries and even the globe. The airline industry, which has been forced by governments in their efforts to curtail COVID-19, has been strategizing to regain its previous overall pattern of an increase annually of airplane flights. The

DOI: 10.4324/9781003202530-8

tourism industry, much of which reliant both upon airplanes and automobiles for transporting tourists across the landscapes of countries and around the globe, is also seeking to reposition in order to achieve business as usual. Manufacturers, such as motor vehicle companies, and the oil industry, also are strategizing how to return to their old ways, even though a sharp drop in manufacturing and transportation of many sorts has resulted in a drop in air pollution and greenhouse gas emissions as the planetary ecosystem and wildlife has undergone at least a temporary reprieved from the damage that global capitalism imposed upon it. In recent months, as new vaccines, most of them developed by Big Pharma resulting in huge profits for its various players, have come on the scene, the cracks in the capitalist world system become once again as the developed countries, including Australia, compete for more than shares against leaving developing countries getting the short end of the stick.

As Prime Minister Scott Morrison hunkers for what he terms a "snap-back" from the impact of the COVID-19 pandemic on the Australian economy, the eco-socialist vision for both humanity and Australia that I set forth in this book may strike many people, perhaps even most climate activists, as completely far-fetched. However, my hope is that as the socio-ecological and climatic crises worsen in the coming decades, that the multitude of humanity, including Australians, will come to the realization that radical action needs to be taken to not only return humanity to a safe climate but a safer place for everyone in terms of access to basic resources essential for a meaningful existence. In short, humanity faces two broad imperatives, how do we learn to live in harmony with each other and how to we learn to live in harmony with nature, as indigenous societies, including in Australia, more or less did effectively for eons prior to the emergence of the archaic state and later the capitalist world system.

I derive hope from the words of Eric Holthaus, a young American journalist, who states:

> In reality, climate change is a symptom of a distorted, imperialist worldview: the continued exploitation of the Earth is justifiable because humans exist separate and can claim over the environment. This worldview, championed by Christians and captains of industries alike, promotes and preserves business as usual and has lead us to the brink of ecological collapse. While a majority of people in the US and Europe [and Australia] accept that climate change is happening, most people reject the need for radical action. Once this mindset changes, however, once we start to see a different world as not only possible but inevitable, all the millions of incremental changes a carbon-free society requires will follow in a whirlwind.
>
> *(Holthaus 2020: 44)*

While there is no guarantee that this will turn out to be the case, to use what many regard to be a cliché, "hope springs eternal." For me, that is what an eco-socialist vision for both humanity and my adopted country of Australia is all about.

Furthermore, given one of the radical measures that I suggest for implementing an eco-socialist vision, namely revisiting and expanding upon public ownership of the means of production and reversing the trend toward privatization fostered in Australia by both ALP and Coalition governments, I am pleased that Holthaus states:

> What about the hundreds of corporations that are responsible for the bulk of emissions? What about the oil and gas industry? We need to nationalize these horrible companies or shut them down. But to do that, we need to talk about it first.
>
> *(Holthaus 2020: 195)*

While Holthaus does not identify himself as either an eco-socialist or an eco-anarchist per se, he is willing to call for radical climate actions. For me, the label is not what is important, even though I use one as a staunch eco-socialist to refer to myself. What is important is to question the existing global political economy, namely capitalism, and propose alternatives for it, ones that envision a new world system based upon social justice, deep democracy, environmental sustainability, and a safe climate.

Returning for a moment to Australia, consider the scenario depicted in George Turner's (1987) eco-novel, *The Sea and the Summer*: it's the 2040s and winter has disappeared as a season in most of the world, the global financial and exchange systems have collapsed, Australia has given up its northern third to accommodate climate refugees from Asia, and the rest of Australia is a police state seeking with an iron hand to maintain a rigid social order. In Melbourne, the "Sweets" strive to maintain some semblance of their privileged lifestyles on higher ground while the abjectly poor "Swills" scramble to survive in decaying high-rise towers surrounded by water. Milner and Burgmann (2021: 364) argue that the "large number of Australian climate fictions might owe something to the legacy of Turner's *The Sea and the Summer*; something to the country's extreme vulnerability to the likely effects of global warming; and something to the, by Anglophone, standards, large Green presence in Australian (and New Zealand) party politics: in 2019 there were 10 elected Green members in the Australian parliament (out of 227) and 8 in the New Zealand (out of 120), as compared to none in the U.S. Congress; 1 (out of 338) in the Canadian parliament, and 1 (out of 650) in the UK." Indeed, the Greens are the most eco-centric and post-materialist party in the Australian federal parliament as well as in various state and territory parliaments, but by and they function largely as green social democrats who I have said on various occasions publicly are "soft on capitalism." Only time will tell whether the Greens can transform themselves into an eco-socialist party. Perhaps the better option would be for eco-socialists and eco-anarchists within the bowels of the Greens to become part of an explicitly anti-capitalist Australian New Left Party.

So let's be honest and clear about what's truly unthinkable, working hard together to make our shared vision of the future our actual future or becoming a not so "lucky country" at the end of the world, much as depicted in Nevil Shute's 1954 classic novel *On the Beach* that portrays the annihilation of the last affluent human beings in the wake of a full-scale nuclear war. The grim realities of ongoing disparities of wealth, climate change, and the distinct possibility of nuclear warfare demand that Australians like Americans and people in the rest of the world contemplate alternatives to the dystopian future which the global capitalist class and its political allies have unleashed.

Achieving an eco-socialist Australia would have to be part of a larger global effort in which a greatly enlarged climate justice movement works in alliance with other anti-systemic movements, such as the global justice, the labor, indigenous and refugee rights, and women's rights movements. Foster and Clark (2020: 271) argue that the ecological revolution will inevitably be a long-term process that may require decades and even centuries. Conversely,

> But given the threat to the earth as a place of human habitation – marked by climate change, ocean acidification, species extinction, loss of freshwater, deforestation, toxic pollution, and more – this process requires immediate reversals in the regime of accumulation. This means opposing the logic of capital, whenever and wherever it seeks to promote the "creative destruction" of the planet. Such a reconstitution of society at large cannot be merely technological, but must transform the human metabolic relation with nature through production and hence the whole realm of social metabolic reproduction.
>
> *(Foster and Clark 2020: 271–272)*

Foster and Clark (2020: 287) argue that ecological revolution will not be smooth but will have to "proceed from one earthly struggle to another," a process that is playing itself out in all countries around the world, including Australia. They argue:

> It is a revolutionary struggle, though, that must commence now with a worldwide movement toward ecosocialism, one capable from its inception of setting limits on capital. This revolt will inevitably find its main impetus in an environmental proletariat, formed by the convergence of economic and ecological crises and the collective resistance of working communities and cultures, a new reality already emerging, particularly in the Global South.
>
> *(Foster and Clark 2020: 287)*

Bearing these thoughts in mind, the Australian climate movement could more fully join in the struggle for social justice both globally and at home if it were to place greater emphasis on climate justice and less on ecological modernization, which has a role to play in climate change mitigation but will not counteract the capitalist treadmill of production and consumption and its drive for continual economic growth. Given the failure of existing international and national climate regimes to date in adequately mitigating climate change, this effort will have to be spurred from below. As I have argued and tried to demonstrate elsewhere, green capitalism and existing climate regimes are not sufficient to mitigate climate change in any serious vein" (Baer 2012). At the other end looms various versions of climate catastrophism which at one level recognize that existing climate regimes and governments are not adopting sufficient measures to mitigate climate change. However, as Labban et al. (2013: 8) argue, climate catastrophism tends to preclude radical solutions to the climate crisis and allows the transnational capitalist class to present "capitalist expansion as the only imaginable, viable solution not only to the problem of climate change, but also to other socio-ecological problems such as food, water, and energy shortages."

Despite efforts in the Soviet Union, China, and numerous other post-revolutionary societies to create socialist systems, all of these efforts were powerfully hindered in achieving this ideal by complex historical, social structural, internal, and external forces. Thus socialism and particularly eco-socialism remains a vision, but one overdue in this era of climate change, environmental devastation, gross social inequality, authoritarianism, militarism, and now COVID-19 and the threat of other pandemics, that humanity faces. Ultimately, the shift to eco-socialism in any country would have to be part of a global process that no one can fully envision at the present time.

Obviously, the transition toward an eco-socialist world is not guaranteed and will require a tedious, even convoluted, path. Nevertheless, while awaiting the "revolution" so to speak, radicals can work on various radical, system-challenging, steps essential to implementing a socio-ecological revolution and ultimately global eco-socialism. The history of the Soviet Union and Stalinism and more recently China, which constitutes a state capitalist society, tell us that socialism cannot be built in one country. The same applies to Australia or any other country in the current era. Although presently or in the near foreseeable future that notion that eco-socialism may be developed in any society, particularly an advanced capitalist one, or blocs of countries may seem utterly ridiculous and utopian, history tells us that social changes, including revolutions, can occur very quickly once economic, political, and social structural changes have reached a tipping point.

Anti-systemic social movements will have to play an instrumental role in bringing about the political which will enable the world to shift to an alternative world system. Given the failure of established international and national climate regimes to date in adequately addressing the climate crisis, meaningful climate action will have to come from below. Ultimately, the climate justice movement will have to form strong alliances with other anti-systemic movements, including the labor, anti-corporate globalization, indigenous and ethnic rights, women's, environmental, and peace movements, in achieving this effort.

While prior to *Climate Crisis and the Global Green New Deal* Noam Chomsky had only fleetingly touched upon climate change in short commentaries, in this book he enters in a series of conversations about climate change with Robert Pollin, a renowned political economist based at the University of Massachusetts. In commenting about the Australian bushfires in late 2019 and early 2020, Chomsky observes that after a climate denialist prime minister "returned grudgingly from a vacation to assure his constituents that he felt their pain," the "opposition labor leader toured the coal plants, calling for expansion of Australia's role as world champion coal exporter and assuring the country that this was quite consistent with Australia's serious comment to combating global warming" (Chomsky and Pollin 2020: 12). Pollin takes William Nordhaus, who received the Nobel Prize in Economics in 2018 for his research on the economics of climate change, to task for being "utterly sanguine about accepting the risks we would face allowing the global mean temperature to rise by 4° C by 2150" (Chomsky and Pollin 2020: 23). In reality, many climate scientists are arguing that if the world does not rapidly reduce greenhouse emissions, a four-degree world would be reached in 2100, not 2150.

Chomsky believes that revival of the labor movement is essential for a variety of reasons, including addressing the environmental crisis, noting that Tony Mazzocchi, head of the Oil, Chemical and Atomic Workers International Union, had been a "harsh critic of capitalism as well as committed environmentalist" (Chomsky and Pollin 2020: 50). While sympathetic to efforts to transcend capitalism, Chomsky asserts that the immediacy of the climate crisis requires addressing it within the "framework of state-capitalist systems," thus his endorsement of a Green New Deal as an interim strategy (Chomsky and Pollin 2020: 58). However, Pollin asserts that a viable Green New Deal would require large-scale public investment, public ownership, and stringent regulation of emissions.

On the matter of eco-socialism, Chomsky does not view it as a viable political project at the present time but admits that it provides a forum for "sharpening ideas" about what a future society might look like (Chomsky and Pollin 2020: 145). Conversely, Pollin asserts that "eco-socialism and the Green New Deal are fundamentally the same project" (Chomsky and Pollin 2020: 146). In contrast to relatively dynamic climate movements in advanced capitalist societies, he argues that they operate at modest levels in most low- and middle-income countries, but concedes that this situation could change quickly.

Going from the present capitalist world system, which has generated and continues to generate greenhouse gas emissions and thereby anthropogenic climate change, to an alternative global political economy, however it is defined, will require much effort, and there are no guarantees that we will be able to create a more socially equitable and environmentally sustainable world. But do we really have any other meaningful choice than to change, or face the downward spiral and the destruction of much of humanity, loss of current biodiversity, and further environmental degradation? Even "green capitalism" does not have the potential to be either socially just or environmentally sustainable, and in light of this reality, humanity must move to something new and better, namely eco-socialism. To paraphrase Rosa Luxemburg, eco-socialism or eco-dystopia.

My own sense, as much as I hate to say it, is that overall things will get worse before they get better, and there is no guarantee that they will get better. Nevertheless, while the capitalist world system appears to be well entrenched on numerous fronts, including within universities, there are numerous cracks in the system, manifested in phenomena such as Brexit in the UK, the election of authoritarian political leaders, various tensions within the European Union, the threat of nuclear war, ongoing conflicts in many parts of the world, growing gaps in global wealth, climatic crises of many sorts, the acidification of the oceans, the extinction of many creatures, and most recently the COVID-19 pandemic.

Presenting a precise timeline for a transition from global capitalism to an eco-socialist world system is extremely difficult, probably impossible. However, the stabilization of the Earth's climate system needs to occur within the next two or three decades lest large swatches of land become uninhabitable for human beings as well as nonhuman species. Despite the daunting difficulties that much of humanity currently faces, it is imperative that eco-socialists keep plugging away at challenging the system in their conversations, teachings, and writings while staying involved in anti-systemic movements by struggling to create new left and anti-capitalist parties, pointing out alternative ways of organizing the world at global, regional, and local levels, and listening to critical input from other progressive perspectives, including eco-anarchist, eco-feminist, de-growth, and Indigenous voices, to mention only a few. In contrast to capitalism, eco-socialist politics is about specifics, production and consumption, employment, a decent wage, deep democracy, dignified conditions of living, public health, and a sustainable and healthy environment.

Perhaps the most important step in the transitional steps delineated about in moving toward eco-socialism is having new left parties assume power, a feat that has occurred with mixed results in places such as Venezuela, Ecuador, and Bolivia under the guise of the Bolivarian Revolution and in Greece with Syriza. Each of these attempts to create a socialism for the 21st century met both internal and external resistance, highlighting the ongoing hegemony of capitalist institutions, be they US imperialism, the European Union, or the International Monetary Fund. Nevertheless, these experiments, despite their shortcomings and the resistance that they encountered, illustrate that the capitalist state can be captured, even if only for a short period of time. As Goodman (2009: 20–21) observes, "for many involved in movements for social change, it is axiomatic that there is no action without hope'. As the existing capitalist world system continues to self-destruct due to its socially unjust and environmentally unsustainable practices, eco-socialism provides a vision to mobilize Australians to prevent their country from becoming even more sunburnt and committed instead to a safe climate and climate justice.

Humanity is obviously at a crossroads, or perhaps more aptly put, at several crossroads: one being business-as-usual; another a shift to some variant of green capitalism that has gained much support globally and in Australians among particularly people somewhere left-of-center; and finally, an eco-socialist vision that while muted at this time will become stronger as the need for it becomes more apparent to the masses of humanity around the world. In the case of Australia, it will take real utopian visions to transform Australia from a sunburnt country in danger of becoming an even more sunburnt place due to climate change to the "lucky country" that Australians, but by no means all, have viewed it as being for some time.

REFERENCES

Adams, Rob. 2009. Transforming Australian cities. In *Climate Changee for Young and Old.* Helen Skyes, ed. Pp. 119–131. Sydney: Future Leaders.

Adams, Tony E., Stacy Holman Jones, and Carolyn Ellis. 2015. *Autoethnography: Understanding Qualitative Research.* Oxford, UK: Oxford University Press.

Albritton, Robert. 2019. *Eco-Socialism for Now and the Future.* Cham, Switzerland: Palgrave Macmillan.

Alexander, Samuel and Brendan Gleeson. 2018. *Degrowth in the Suburbs: A Radical Urban Imagery.* Singapore: Palgrave Macmillan.

_____. 2020. *Urban Awakenings: Disturbance and Enchantment in the Industrial City.* Singapore: Palgrave Macmillan.

Altman, Dennis. 2006. *51st State?* Melbourne: Scribe Short Books.

Altman, Jon. 2009. Indigenous communities: Miners and the state in Australia. In *Power, Culture, Economy: Indigenous Australians and Mining.* Jon Altman and David Martin, eds. Pp. 17–48. Canberra: ANU Press.

Altman, Jon and Sean Kerins. 2012. *People on Country: Vital Landscape/Indigenous Futures.* Sydney: Federation Press.

Amin, Samir. 1990. *Delinking: Towards a Polycentric World.* London: Zed Books.

_____. 2013. *The Implosion of Contemporary Capitalism.* New York: Monthly Review Press.

Amos, C. and T. Swanm. 2015. Carmichael in context: Quantifying Australia's threat to climate action. Australia Institute Discussion paper. http://www.tai.org.au/context/carmichael-context.

Anderson, James. 2006. Afterword: Only sustain . . . the environment, "anti-globalization," and the runaway bicycle. In *Nature Revenge: Reclaiming Sustainability in an Age of Corporate Globalization.* Josse Johnston, Michael Gismondi, and James Goodman, eds. Pp. 245–273. Toronto: Broadview.

Angus, Ian. 2016. *Facing the Anthropocene, Fossil Capitalism and the Crisis of the World System.* New York: Monthly Review Press.

_____. 2017. *A Redder Shade of Green: Intersections of Science and Socialism.* New York: Monthly Review Press.

Arboleda, Mark. 2020. *Planetary Mine: Territories of Extraction Under Late Capitalism.* London: Verso.

Armstrong, Mick and Tom Bramble. 2007. *The Labor Party: A Marxist Analysis*. Melbourne: Socialist Alternative.

ATA Energy Projects Team. 2017.100% renewable grid by 2030: Discussion paper. November. Melbourne: Alternative Technology Association.

Aulby, Hannah. 2017. Undermining our democracy: Foreign influence through the Australian mining lobby. Discussion paper. Australia Institute, August. www.tai.org.au, accessed July 8, 2019.

Aulby, Hannah and Mark Ogge. 2016. Greasing the wheels: The systemic weaknesses that allow undue influence by mining companies on government: A Queensland case study. Australia Institute, July. www.tai.org.au, accessed July 8, 2019.

Australian Conservation Foundation. 2015. Australia's top 10 climate polluters. http://www.acf@acfonline.org.au.

_____. 2020. Fossil fuel money: Distorting democracy. www.acf@acfonline.org.au.

_____. n.d. License to kill: Commonwealth environmental approval for Adani's Carmichael coal mine project. http://www.acf@acfonline.org.au.

Australian Council of Trade Unions. 1992. *The Greenhouse Effect. Employment & Development*. Melbourne: ACTU.

Australian Government. 2012. *Energy White Paper*. Canberra, ACT: Australian Government.

_____. 2018. Electricity markets and the role of coal fired power. September. www.aph.gov.au/Parliamentary_Business/Committees/Senate, accessed January 13, 2021.

Australian Government, Department of Industry, Science, Energy and Resources. 2021. Quarterly update of Australia's national greenhouse gas inventory: September 2020. www.industry.gov.au, accessed March 10, 2021.

Australian Greens. 2015. Renew Australia: Power the new economy. http://www.green.org.au/renew, accessed September 27, 2019.

_____. 2018. Economic justice. November. http://www.greens.org.au/policies/economic-justice.

_____. 2020a. Renew Australia 2030. www.green.org.au, accessed September 27, 2020.

_____. 2020b. Towards a Green new deal. www.green.org.au, accessed January 3, 2021.

Australia Institute. 2016. Climate of the nation 2016: Australian attitudes on climate change. Australia Institute, June. http://www.australiainstitute.com.au, accessed March 23, 2018.

_____. 2017. Polling shows voters don't want the Adani mine. http:// www.tai.org.au, accessed october 8, 2017.

_____. 2018. Polling – CEO salaries, May. http//www.tai.org.au, accessed December 15, 2019.

Australian Petroleum Production & Exploration Association. 2020. Key statistics 2020. Australian Petroleum Production & Exploration Association Limited. www.appea.com.au, accessed March 15, 2021.

Baer, Hans A. 2012. *Global Capitalism and Climate Change: The Need for an Alternative World System*. Lanham, MD: AltaMira Press.

_____. 2016. Toward democratic eco-socialism as the next world system. The Next System Project. April 28. http://www.thenextsystem.org/learn/collections/new-system-possibilities-and-proposals.

_____. 2018. *Democratic Eco-Socialism as a Real Utopia: Transitioning to an Alternative World System*. New York: Berghahn.

_____. 2019. The larger struggling: Mitigating capitalism – A contribution to the GTI forum. *The Climate Movement: What's Next?* June. *Great Transition Initiative: Toward a Transformative Vision and Praxis*. http//www.greattransition.org.

_____. 2021. *Grappling with Societies and Institutions in an Era of Socio-Ecological Crisis: Journey of a Radical Anthropologist*. Lanham, MD: Lexington Books.

Baer, Hans A. and Merrill Singer. 2009. *Global Warning and the Political Ecology of Health*. Walnut Creek, CA: Left Coast Press.

_____. 2014. *The Anthropology of Climate Change: An Integrated Critical Perspective*. London: Earthscan at Routledge.

_____. 2018. *The Anthropology of Climate Change: An Integrated Critical Perspective* (2nd Edition). London: Earthscan at Routledge.

Baer, Hans A., Merrill Singer, and Ida Susser. 1997. *Medical Anthropology and the World System: A Critical Perspective*. South Hadley, MA: Bergin & Garvey.

_____. 2003. *Medical Anthropology and the World System: A Critical Perspective* (2nd Edition). Westport, CT: Praeger.

_____. 2013. *Medical Anthropology and the World System: A Critical Perspective* (3rd Edition). Westport, CT: Praeger.

Bandt, Adam. 2016. Making progressive government happen. In *How to Move Progressive in Australia: Labor or Green?* Dennis Altman and Sean Scalmer, eds. Pp. 179–202. Melbourne: Monash University Publishing.

Barber, Marcus. 2011. Talking about the weather: Anthropology and climate change. Paper presented at the International Union of Anthropological and Ethnological Sciences/Australian Anthropological Society/Association of Social Anthropologists of Aotearora New Zealand Joint Conference, University of Western Australia, Perth, July 4–8.

Barnes, Alison. 2020. Scorched summer reminds us: Climate change is union business. *NTEU Advocate* 27(1): 2.

Bendell, Jem. 2018. Deep Adaptation: A map for navigating climate tragedy. IFLAS occasional paper 2. http://lifeworth.com/deepadaptation.pdf, accessed October 21, 2019.

Bennett, Ebony. 2018. Climate of the nation 2018: Tracking Australia's attitudes towards climate change and energy. The Australia Institute, September.

Benns, M. 2011. *Dirty Money: The True Cost of Australia's Mineral Boom*. Sydney: William Heinemann.

Beresford, Quentin. 2018. *Adani and the War Over Coal*. Sydney: New South.

Beyond Zero Emissions, the Melbourne Energy Institute and the German Aerospace Centre 2014. *Zero Carbon Australia: High Speed Rail*. Melbourne: Beyond Zero Emissions, the Melbourne Energy Institute and the German Aerospace Centre.

Birch, Tony. 2016. Climate change, recognition and social place-making. In *Unstable Relations: Indigenous People and Environmentalism in Contemporary Australia*. Eve Vincent and Timothy Neale, eds. Pp. 356–383. Perth: University of Western Australia Publishing.

Bodley, John H. 2017. *Cultural Anthropology: Tribes, States, and the Global System* (6th Edition). Lanham, MD: Rowman & Littlefield.

Bond, Patrick. 2012. *Politics of Climate Justice: Paralysis Above, Movement Below*. Scottsville, South Africa: University of KwaZulu-Natal Press.

Bookchin, Murray. 1991. *The Ecology of Freedom* (2nd Edition). Montreal: Black Rose Books.

Boswell, Terry and Christopher Chase-Dunn. 2000. *The Spiral of Capitalism and Socialism*. Boulder, CO: Lynne Rienner.

Boyce, James. 2019. The devil and Scott Morrison: What do we know about the prime minister's Pentecostalism. *The Monthly*, February.

Braganza, K., K. Hennessey, L. Alexander, and B. Trewin. 2014. Changes in extreme weather. In *Four Degrees of Global Warming: Australia in a Hot World*. Peter Christoff, ed. Pp. 33–59. New York: Routledge.

Bramble, Tom. 2005. Labour movement leadership. In *Class and Struggle in Australia*. Rick Kuhn, ed. Pp. 74–89. Sydney: Pearson Education Australia.

Bramble, Tom and Rick Kuhn. 2010. *Labor's Conflict: Big Business, Workers and the Politics of Class*. Melbourne: Cambridge University Press.

Breakthrough. n.d. www.breakthroughonline.org.au, accessed January 3, 2021.

Brett, Judith. 2020. The coal curse: Resources, climate and Australia's future. *Quarterly Essay*, No. 78: 1–81.

Brohe, Arnaud, Nick Eyre, and Nicholas Howarth. 2009. *Carbon Markets: An International Business Guide*. London: Earthscan.

Broughton, Alan and Elena Garcia. 2017. *Sustainable Agriculture versus Corporate Greed: Small Farmers, Food Security & Big Business*. Sydney: Resistance Books.

Broughton, Alan and Elena Garcia. 2021. Nationals don't speak for farmers on climate change action. *Green Left*, February 16.

Brown, Bob. 2004. *Memo for a Saner World*. Melbourne: Penguin.

Brown, Bob and Peter Singer. 1996. *The Greens*. Melbourne: Text Publishing.

Brown, Lester R. 2009. *Plan B 4.0: Mobilizing to Save Civilization*. New York: W.W. Norton & Company.

Bruckner, M., A. Durey, R. Mayes, and C. Pforr. 2014. Confronting the 'resource curse or cure'. In *The Sustainability of Development of Development in Western Australia*. M. Bruckner, A. Durey, R. Mayes, and C. Pforr, eds. Pp. 3–23. New York: Springer.

Burgmann, Meredith and Verity Burgmann. 1998. *Green Bans, Red Unions: Environmental Action and the New South Wales Builders' Federation*. Sydney: UNSW Press.

Burgmann, Verity. 2003. *Power, Profit and Protest: Australian Social Movements and Globalisation*. Sydney: Allen & Unwin.

————. 2013. From 'jobs versus environment' to 'green-collar jobs: Australian trade unions and the climate change debate. In *Trade Unions in the Green Economy: Working for the Environment*. Nora Raethzel and David Uzzell, eds. Pp. 131–145. London: Earthscan from Routledge.

Burgmann, Verity and Hans A. Baer. 2012. *Climate Politics and the Climate Movement in Australia*. Melbourne: Melbourne University Press.

Butler, Mark. 2017. *Climate Wars*. Melbourne: Melbourne University Press.

Cahill, Damien and Stephen Brown. 2008. The rise and fall of the Australian Greens: The 2002 Cunningham by-election and its implications. *Australian Journal of Political Science* 43: 259–275.

Campbell, Rod, Eliza Littleton, and Alia Armistead. 2021. Federal and state government assistance to fossil fuel producers and major users. Australia Institute, April, http://www.tai.org.au.

Carey, Adam, Anna Prytz, and Madeline Hefferman. 2020. Bursting the foreign student bubble puts squeeze on research. *The Age*, June 6, p. 8.

Catley, Bob. 1996. *Globalising Australian Capitalism*. Melbourne: Cambridge University Press.

CFMEU. 2007. Climate change position paper. September. http//www.cfmeu.com.au.

Chancel, Lucas. 2020. *Unsustainable Inequalities: Social Justice and the Environment*. Cambridge, MA: Belknap Press of Harvard University Press.

Chase-Dunn, Christopher. 2005. Social evolution and the future of world society. *Journal of World-Systems Research* 11(2): 171–192.

Chomsky, Noam and Robert Pollin. 2020. *Climate Crisis and the Global New Deal*. London: Verso.

Christoff, Peter. 2014. Four degrees or more? In *Four Degrees of Global Warming: Australia in a Hot World*. Peter Christoff, ed. Pp. 1–14. London: Earthscan from Routledge.

Chubb, Philip. 2014. *Power Failure: The Inside Story of Climate Politics Under Rudd and Gillard*. Melbourne: Black, Inc.

Cleary, Paul. 1991. How we can fight our way out of the greenhouse. *Sydney Morning Herald Magazine*, October 26, p. 1.

———. 2011. *Too Much Luck: The Mining Boom and Australia's Future*. Melbourne: Black, Inc.

———. 2012. *Too Much Luck: The Mining Boom and Australia's Future*. Melbourne: Black, Inc.

Climate Council. 2015. Galilee Basin – unburnable coal. www.climatecouncil.org.au, accessed February 22, 2016.

———. 2016a. Climate council season update; Abnormal 2016. www.climatecouncil.org.au, accessed April 27, 2017.

———. 2016b. Renewable energy jobs: Future growth in Australia. http://www.climate-council.org.au, accessed April 27, 2017.

———. 2019a. Climate cuts, cover-ups and censorship. www.climatecouncil.org.au, accessed December 5, 2019.

———. 2019b. Compound costs: How climate change is damaging Australia's economy. www.climatecouncil.org.au, accessed May 16, 2020.

———. 2020. Tropical cyclones and climate change: Factsheet. http://www.climatecouncil.org.au, accessed December 10, 2020.

Climate Institute. 2016. Climate of the nation 2016: Australian attitudes on climate change. http://www.australiainstitute.org.au.

Collins, Paul. 2009. *Burn: The Epic Story of Bushfire in Australia*. Melbourne: Scribe.

Commonwealth Science and Industrial Resource Organisation. 2015. *Climate Change in Australia: Projections for Selected Australian Cities*. Canberra: Department of the Environment and Bureau of Meteorology.

Connor, Linda H. 2016. *Climate Change and Anthropos: Planet, People and Places*. London: Routledge.

Connor, Linda, Sonia Freeman, and Nick Higgenbotham. 2009. Not just a coalmine: Shifting grounds of community opposition to coal mining in southeastern Australia. *Ethnos* 74: 490–513.

Cooke, Sophie. 2010. Leave it in the ground – the growing global struggle against coal. In *Sparking a Worldwide Energy Revolution: Social Struggles in the Transition to a Post-Petrol World*. Kolya Abraamsky, ed. Pp. 424–438. Oakland, CA: AK Press.

Corrighan, T. 1980. The political economy of minerals. *Journal of Australian Political Economy* 7: 28-40.

Cooper, Liam and Hans A. Baer. 2019. *Urban Eco-Communities in Australia: Real Utopian Responses to the Ecological Crisis or Niche Markets?* Singapore: Springer.

Crough, G.J., and E.L. Wheelwright. 1983. Australia: Client state of international capital: A case study of the mineral industry. In *Essays in the Political Economy of Australian Capitalism*. E.L. Wheelwright and K. Buckley, eds. Pp. 15–42. Sydney: Australia and New Zealand Book Company.

Crough, Greg and Ted Wheelwright. 1982. *Australia: Client State*. Melbourne: Penguin Books Australia.

Crowe, David. 2019. *Venon: Vendettas, Betrayals and the Price of Power*. Sydney: Harper Collins.

Crowe, Shaun. 2018. *Whitlam's Children: Labor and Greens in Australia*. Melbourne: Melbourne University Press.

Commonwealth Scientific and Industrial Research Organisation and Australian Bureau of Meteorology. 2020. *State of the Climate 2020*. Melbourne: CSIRO and Australian Bureau of Meteorology. www.bing.com/search?q=csiro+-+state+of+the+climate+2020&cvid=d3d20db4068046cb92b98d8340c07858&aqs=edge..69i57j0.19070j0j4&FORM=ANAB01&DAF0=1&PC=U531, accessed February 27, 2021.

Crowley, Kate. 1999. Jobs and environment: The "double dividend" of ecological modernisation? *International Journal of Social Economics,* No. 26: 1013–1026.

Cunningham, Michelle, Luke Van Uffelen, and Mark Chambers. 2019. The changing global market for Australian coal. *Reserve Bank of Australia,* Pp. 28–38.

Cunningham, Michelle, Luke Nanuffelen, and Mark Chambers. The changing global market for Australian coal. *Reserve Bank of the Australia – Bulletin,* September 19, http:www.rba.gov.au/bulletin/2019/sep/

Dahlgren, Kari. 2019. Digging deeper: Precarious futures in two Australian coal mining towns. PhD thesis, Department of Anthropology, London School of Economics, August.

Davidson, S. and A. de Silva. 2013. The Australian coal industry – adding value to the Australian economy. Report prepared for the Australian Coal Association, April.

Davis, Mark. 2008. *The Land of Plenty: Australia in the 2000s.* Melbourne: Melbourne University Press.

Davison, Graeme. 2004. *Car Wars: How the Car Won the Hearts and Conquered the Cities.* Sydney: Allen & Unwin.

Dawson, Ashley. 2017. *Extreme Cities: The Peril and Promise of Urban Life in the Age of Climate Change.* London: Verso.

Day, Norman. 2020. Unbuilding cities as high-rises reach their use-by date. In *The Year That Changed Us.* Molly Glassey, ed. Pp. 118–122. Melbourne: Thanes & Hudson.

Denniss, Richard. 2016. *Econobabble: How to Decode Political Spin and Economic Nonsense.* Melbourne: Black, Inc.

———. 2017. *Curing Affluenza: How to Buy Less Stuff and Dave the World.* Melbourne: Black, Inc.

———. 2018. How neoliberalism ate itself and what comes next? *Quarterly Essay,* No. 70: 1–79.

———. 2019. *Dead Right: How Neoliberalism Ate Itself and What Comes Next.* Melbourne: Black, Inc.

Department of Infrastructure and Regional Development. 2017. Managing the carbon footprint of Australian aviation. August. Canberra: Australian Government. http://www.infracture.gov.au, accessed September 2, 2019.

Department of Resources and Energy. 1986. *Energy 2000 – A National Review.* Canberra, ACT: Australian Government.

Derber, Charles. 2010. *Greed to Green: Solving Climate Change and Remaking the Economy.* Boulder, CO: Paradigm Publishers.

Diamond, Jared. 2005. *Collapse: How Societies Choose to Fail or Survive.* Melbourne: Penguin/Allen Lane.

Diesendorf, Mark. 2009. *Climate Action: A Campaign Manual for Greenhouse Solutions.* Sydney: UNSW Press.

Doherty, Brian and Timothy Doyle. 2018. Friends of the Earth International: Agonistic politics, modus vivendi and political change. *Environmental Politics* 27: 1057–1078.

Doig, Tom. 2019. *Hazelwood.* Melbourne: Penguin/Viking.

Dominick, Brian. 2008. From here to Parecon: Thoughts on Strategy for Economic Revolution. In *Real Utopia: Participatory Society for the 21st Century.* C. Spannos, ed. Pp. 380–395. Oakland, CA: AK Press.

Earthworker. n.d. http://earthworkercooperative.com.au/our-mission/, accessed March 27, 2019.

Ede, Andrew. 2019. *Technology and Society: A World History.* Cambridge, UK: Cambridge University Press.

Edwards, Lindy. 2020. *Corporate Power in Australia: Do the 1% Rule?* Melbourne: Monash University Publishing.

Energy Minerals Branch. 1999. *Australia's Export Coal Industry*. Canberra, ACT: Australian Government, Department of Industry, Sciences and Resources.

Eriksen, Thomas Hylland. 2016. *Overheating: An Anthropology of Accelerated Change*. London: Pluto Press.

Evans, Geoff. 2010/2011. A rising tide: Linking local and global climate justice. *Journal of Australian Political Economy*, No. 66: 198–221.

Evans, Michael. 2019. Racing towards a climate catastrophe. *NTEU Advocate* 26(2): 26–27.

Falk, Jim and Andrew Brownlow. 1989. *The Greenhouse Challenge: What's to be Done?* Melbourne: Penguin Books.

Farstad, Fay Madeleine. 2019. Does size matter? Comparing the party politics of climate change in Australia and Norway. *Environmental Politics* 28: 997–1016.

Fernandes, Clinton. 2018. *Island off the Coast of Asia: Instruments of Statecraft in Australian Foreign Policy*. Melbourne: Monash University Press.

Flannery, Tim. 2005. *The Weather Makers*. New York: Atlantic Monthly Press.

———. 2020. *The Climate Cure: Solving the Climate Emergency in the Era of COVID-19*. Melbourne: Text Publishing.

Foley, Gary. 2007. The Australian labor party and the *Native Title Act*. In *Sovereign Subjects: Indigenous Sovereignty Matters*. A. Moreton-Robinson, ed. Pp. 118–139. Sydney: Allen & Unwin.

Foster, John Bellamy. 2009. *The Ecological Revolution: Making Peace with the Planet*. New York: Monthly Review Press.

———. 2020a. Ecology. In *The Marx Revival: Key Concepts and New Interpretations*. Marcello Musto, ed. Pp. 177–196. Cambridge, UK: Cambridge University Press.

———. 2020b. The renewal of the socialist ideal. *Monthly Review* 74(4): 1–13.

Foster, John Bellamy and Brett Clark. 2020. *The Robbery of Nature: Capitalism and the Ecological Rift*. New York: Monthly Review Press.

Frank, Christine. 2009. The bankruptcy of capitalist solutions to the climate crisis. *Capitalism Nature Socialism* 20: 32–43.

Frankel, Boris. 1992. *From the Prophets Deserts Come: The Struggle to Reshape Australian Political Culture*. Melbourne: Arena Publications.

———. 2001. *When the Boat Comes In: Transforming Australia in the Age of Globalisation*. Sydney: Pluto Australia.

———. 2018. *Fictions of Sustainability: The Politics of Growth and Post-Capitalist Futures*. Melbourne: Greenmeadows.

———. 2020. *Capitalism versus Democracy? Rethinking Politics in the Environmental Crisis*. Melbourne: Greenmeadows.

Fredman, Nick. 2013. Watermelons or tomatoes? Social democracy, class and the Australian Greens. *Capitalism Nature Socialism* 24(4): 86–104.

Fredman, Nick, Pip Hinman, and Susan Price. 2013. Revolutionary unity to meet the capitalist crisis. *Marxist Left Review*, No. 6: 57–77.

Friends of the Earth Melbourne. 2019. Transforming VIC: Creating jobs while cutting emissions. 4 July version.

Garnaut, Ross. 2019. *Super-Power: Australia's Low-Carbon Opportunity*. Melbourne: La Trobe University Press.

———. 2021. *Reset: Restoring Australia after the Pandemic Recession*. Melbourne: La Trobe University Press.

Geoscience. 2020. Coal. http://www.ga.gov.au, accessed January 12, 2021.

Gergis, Joelle. 2018. *Sunburnt Country: The History and Future of Climate Change in Australia.* Melbourne: Melbourne University Press.

Gerster, Robin and Jan Basnett. 1991. *Seizures of Youth: The Sixties and Australia.* Melbourne: Hyland House.

Gilding, Paul. 2011. *The Great Disruption: How the Climate Crisis Will Transform the Global Economy.* London: Bloomsbury.

———. 2019. *Climate Emergency Defined and Does the Evidence Justify One.* Melbourne: Breakthrough National Centre for Climate Restoration.

Gilfillian, Geoff and Chris McGann. 2018. Trends in union membership in Australia. Parliamentary Library: Statistical Snapshot. October 15.

Gleeson, Brendan. 2010. *Lifeboat Cities.* Sydney: UNSW Press.

Gleeson, Margaret. 2018, article appeared in Green Left, November 8. Under pressure, Adani scales down its coalmine project.

Goldie, Cassandra and Peter Saunders. 2018. *Inequality in Australia 2018.* Sydney: Australian Council of Social Services and University of New South Wales.

Gomeroi Tribal Nation Secretariat. 2012. htpp://www.environment.nsw.gov.au/resources/legislation/GomeroiTNSsub.pdf.

Goodman, James. 2009. Climate movement: A new politics of moral protest? *Chain Reaction,* (17): 22–24.

Goodman, James and D. Worth. 2008. The minerals boom and Australia's 'resource curse'. *Journal of Australian Political Economy,* No. 61: 201–209.

Goods, Caleb. 2011. Labour unions, the environment and 'green jobs'. *Journal of Australia Political Economy,* No. 67: 47–67.

Gorz, Andre. 1980. *Ecology as Politics.* Boston: South End Press.

Graham, Stephen. 2016. *Vertical: The City from Satellites from Bunkers.* London: Verso.

Grattan, Michelle. 2020. Kevin Rudd. In *Australian Prime Ministers.* Michelle Grattan, ed. Pp. 469–489. Sydney: New Holland.

Grattan, Michelle and Jane Seaborn. 2019. The independents: Cases of Warringah and Wentworth. In *From Turnbull to Morrison: The Trust Divide.* Mark Evans, Michelle Grattan, and Brendan McCaffrie, eds. Pp. 281–294. Melbourne: Melbourne University Press.

Green, Donna. 2009. Opal waters, rising sea: How sociocultural inequality reduces resilience. In *Anthropology and Climate Change: From Encounters to Action.* Susan A. Crate and Mark Nuttall, eds. Pp. 218–227. Walnut Creek, CA: Left Coast Press.

Green, Donna, Jack Billy, and Alo Tapim. 2010. Indigenous Australians' knowledge of weather and climate. *Climatic Change,* No. 100: 337–354.

Green, Donna and Liz Minchin. 2010. *Screw Light Bulbs: Smarter Ways to Save Australians Time and Money.* Perth: University of Western Australia Press.

———. 2014. Living on climate-changed country: Indigenous health, well-being and climate change in remote Australian communities. *EcoHealth* 11: 266–274.

Greenland, Hal. 2017. The Greens – a house of many mansions. *Green Left,* January 31.

Greenpeace Australia Pacific. 2012a. Cooking the climate, wrecking the reef: The global impact of coal exports from Australia. Sydney: Greenpeace Australia Pacific. http://www.greenpeace.org.au, accessed September 23, 2018.

———. 2012b. Boom goes the Reef: Australia's coal export boom and the industrialisation of the Great Barrier Reef. Sydney: Greenpeace Australia Pacific. http://www.greenpeace.org.au, accessed September 23, 2018.

———. 2014. Activists halt Whitehaven's gentle bulldozers. Press release. May 29, http://www. Greenpeace.org.au/news.

———. 2020. Recover and prosper: Recharging the Australian economy with clean energy. http://www.greenpeace.org.au, accessed September 23, 2018.

Gregorie, Paul. 2019. Climate crisis is an 'extraordinary emergency' that merits breaking the law. *Green Left*, September 17.

Grubert, Emily and Whitney Skinner. 2017. A town divided: Community values and attitudes towards coal seam gas development in Gloucester, Australia. *Energy Research & Social Science* 30: 43–52.

Grudnoff, Matt. 2018. Gini out of the bottle. Discussion paper. Australia Institute, June.

———. 2020. The carbon pricing mechanism under the Gillard government. Australia Institute, August.

Habibis, Daphne and Maggie Walter. 2015. *Social Inequality in Australia: Discourses, Realities and Futures* (2nd Edition). Melbourne: Oxford University Press.

Hallam, Roger. 2019. *Common Sense for the 21st Century: Only Nonviolent Rebellion Can Now Stop Climate Breakdown and Social Collapse*. White River Junction, VT: Chelsea Publishing.

Halliday, Fred. 1999. *Revolution and World Politics: The Rise and Fall of the Sixth Great Power*. Durham, NC: Duke University Press.

Hamilton, Clive. 2001. *Running from the Storm: The Development of Climate Change Policy in Australia*. Sydney: UNSW Press.

———. 2007. *Scorcher: The Dirty Politics of Climate Change*. Melbourne: Black, Inc.

———. 2010. *Requiem for a Species: Why We Resist the Truth about Climate Change*. Sydney: Allen & Unwin.

———. 2018. *Silent Invasion: China's Influence in Australia*. Melbourne: Hardie Grant Books.

Hamilton, Clive and Richard Denniss. 2005. *Affluenza: When Too Much Is Never Enough*. Sydney: Allen & Unwin.

Hamilton, Clive and Mareike Ohlberg. 2020. *Hidden Hand: Exposing How the Chinese Communist Party Is Reshaping the World*. Melbourne: Hardie Grant Books.

Hammond, John L. 2012. Social movements and struggles for socialism. In *Taking Socialism Seriously*. Anatole Anton and Richard Schmitt, eds. Pp. 213–247. Lanham, MD: Lexington Books.

Hardt, Michael and Antonio Negri. 2009. *Commonwealth*. Cambridge, MA: Harvard University Press.

Hastings, Graham. 2005. Students in classes. In *Class and Struggle in Australia*. Rick Kuhn, ed. Pp. 92–106. Sydney: Pearson Education Australia.

Hawke, Robert J. 1989. Our country, our future (State on the Environment). Canberra: Commonwealth of Australia, AGPS, https://pmtranscripts.pmc.gov.au/sites/default/files/original/0000766 . . . , accessed December 5, 2020.

Helm, Dieter. 2017. *Burn Out: The Endgame for Fossil Fuels*. New Haven, CT: Yale University Press.

Henderson-Sellers, Ann and Richard Blong. 1989. *The Greenhouse Effect: Living in a Warmer World*. Sydney: NSW University Press.

Henzel, Ted. 2007. *Australian Agriculture: Its History and Challenges*. Melbourne: CSIO Press.

Hepburn, John, Bob Burton, and Sam Hardy. 2010. Stopping the Australian coal export boom. Greenpeace Australia. http://wwww. KM_754e-20160524192153 (parliament.qld.gov.au).

Hero Project. 2009. *Nukkan Kungun Yunnan – Ngarrindjeri's Being Heard*. www.youtube.com/watch?v+dSsv-dSS40, accessed May 23, 2019.

Hickel, Jason. 2020. *Less Is More: How Degrowth Will Save the World*. London: William Heinemann.

Hilier, Ben. 2010. A Marxist critique of the Australian Greens. *Marxist Left Review*, No. 1: 3–35.

Hinman, Pip. 2019. Labor's great gas con. *Green Left*, May 7.

Holmes, Dave. 2021. Putting the anti-China campaign in perspective. *Green Left*, March 2, p. 10.

Holmes, David. 2012. Changing the climate: Modernity at its limits. In *Australian Sociology: A Changing Society*. David Holmes, Kate Hughes, and Roberta Jillian, eds. Pp. 386–406. Frenchs Forest NSW: Pearson Australia.

Holthaus, Eric. 2020. *The Future Earth: A Radical Vision for What's Possible in the Age of Warming*. New York: HarperOne.

Hornborg, Alf. 2001. *The Power of the Machine: Global Inequalities of Economy, Technology, and Environment*. Walnut Creek, CA: AltaMira Press.

Horne, Donald. 1964. *The Lucky Country*. Melbourne: Dent.

Hudson, Marc. 2019. 'A form of madness': Australian climate and energy policies 2009–2018. *Environmental Politics* 28: 583–589.

Hulme, Mike. 2013. *Exploring Climate Change Through Science and in Society: An Anthology of Mike Hulme's Essays, Interviews and Speechs*. New York: Earthscan at Routledge.

Humphreys, Jordan. 2019. The political economy of immigration in Australia. *Marxist Left Review*, No. 17: 29–56.

Hutchings, Suzi. 2019. Indigenous anthropologists caught in the middle: The fragmentation of indigenous knowledge in native title anthropology, law, and policy in urban and rural Australia. In *Transcontinental Dialogues: Activist Alliances with Indigenous Peoples of Canada, Mexico, and Australia*. R. Aida Hernandez Castillo, Suzi Hutchings, and Brian Noble, eds. Pp. 193–219. Tucson: University of Arizona Press.

Hutton, Drew and Libby Connors. 2004. The left dilemma for the Greens. *Social Alternatives* 23(1): 34–37.

International Energy Agency. 2018. Energy policies of IEA countries: Australia 2018 review. International Energy Agency. www.iea.org

Jackson, Stewart. 2016. *The Australian Greens: From Activism to Australia's Third Party*. Melbourne: Melbourne University Press.

Jackson, Tim. 2009. *Prosperity without Growth*. London: Earthscan.

Jones, Barry. 2020. *What Is to Be Done? Political Engagement and Saving the Planet*. Melbourne: Scribe.

Joshi, Ketan. 2020. *Windfall: Unlocking a Fossil-Free Future*. Sydney: NewSouth.

Jowit, Juliette and Patrick Wintour. 2008. Cost of tackling global climate change has doubled, warns Stern. *The Guardian*, June 26, p. 1.

Kaitsu, Moeno, Tom Swann, and Audrey Quickie. 2019. Hytrojan: Is hydrogen the next "clean coal"? Australia Institute, October.

Kallis, Giorgos. 2018. *Degrowth*. Newcastle upon Tyne, UK: Agenda Publishing.

Kelly, Dominic. 2019. *Political Troglodytes and Economic Lunatics: The Hard Right in Australia*. Melbourne: La Trobe University Press.

Klein, Naomi. 2014. *This Changes Everything: Capitalism vs. the Climate*. London: Allen Lane.

_____. 2019. *On Fire: The Burning Case for a Green New Deal*. London: Allen Lane.

Kolaski, Erald. 2021. The ecological state. *Monthly Review* 72(8): 23–36.

Kolbert, Elizabeth. 2014. *The Sixth Extinction: An Unnatural History*. New York: Henry Holt and Company.

Kovel, Joel and Saul Quincey. 2019. *The Emergence of Ecosocialism: Collected Essays by Joel Kovel*. New York: 2 Leaf Press.

Krien, Anna. 2017. Coal, coral and Australia's climate deadlock. *Quarterly Essay*, No. 66: 1–116.

Labban, Mazen, David Coreia, and Matt Huber. 2013. Apocalypse, the radical left and the post-colonial condition. *Capitalism, Nature, Socialism* 24(1): 6–8.

Lancaster, Steve. 2012. *Green Australia: A Snapshot.* Kent Town, South Australia: Wakefield Press.

Langness, L.L. and Gelya Frank. 1981. *Lives: An Anthropological Approach to Biography.* Novato, CA: Chandler and Sharp.

Langton, Marcia. 2011. Anthropology, politics and the changing world of Aboriginal Australians. *Anthropological Forum* 21: 1–22.

———. 2013. *Boyer Lectures 2012: The Quiet Revolution: Indigenous People and the Resources Boom.* Sydney: ABC Books.

Langton, Marcia and Odette Mazel. 2012. The resource curse compared: Australian Aboriginal participation in the resource extraction industry and distribution of impacts. In *Community Futures, Legal Architecture: Foundations for Indigenous Peoples in the Global Mining Boom.* Marcia Langton and Judy Longbottom, eds. Pp. 23–44. London: Routledge.

Leahy, Terry. 2011. The gift economy. In *Life without Money: Building Fair and Sustainable Economics.* Anitra Nelson and Franz Timmerman, eds. Pp. 111–135. London: Pluto Press.

Lee, M. and S. Draper. 1988. The coal industry: The current crisis and the campaign for a National Coal Authority. *Journal of Australian Political Economy*, No. 23: 45–60.

Liegey, Vincent and Anitra Nelson. 2021. *Exploring Degrowth: A Critical Guide.* London: Pluto Press.

Lin, B. and X. Li. 2011. The effect of carbon tax on per capita CO_2 emissions. *Energy Policy* 39: 5137–5146.

Lock the Gate. n.d. http://www.lockthegate.org.au/groups.

Loewenstein, Antony. 2013. *Profits of Doom: How Vulture Capitalism Is Swallowing the World.* Melbourne: Melbourne University Press.

Loewy, Michael. 2015. *Ecosocialism: A Radical Alternative to Capitalist Catastrophe.* Chicago: Haymarket Books.

———. 2020. Revolution. In *The Marx Revival: Key Concepts and New Interpretations.* Marcello Musto, ed. Pp. 126–140. Cambridge, UK: Cambridge University Press.

Lohmann, Gui and Jacob Trischler. 2017. License to build, license to change? Market power, pricing and financing of airport infrastructure development in Australia. *Transport Policy* 59: 28–37.

Lovelock, James. 2009. *The Vanishing Face of Gaia: The Life of an Independent Scholar.* London: Souvenir Press.

Low, Nicholas. 2013. From automobility to sustainable transport. In *Transforming Urban Transport: The Ethics, Politics and Practices of Sustainable Mobility.* Nicholas Low, ed. Pp. 57–58. London: Earthscan.

Lowe, Ian. 1989. *Living in the Greenhouse.* Newham, VIC: Scribe Publications.

———. 2016. *The Lucky Country? Reinventing Australia.* Brisbane: University of Queensland Press.

Lynas, Mark. 2020. *Our Final Warning: Six Degrees of Climate Emergency.* London: 4th Estate.

Macintyre, Stuart. 2016. *A Concise History of Australia* (4th Edition). Cambridge, UK: Cambridge University Press.

Mackay, Hugh. 2018. *Australia Reimagined: Towards a More Compassionate, Less Anxious Society.* Sydney: Pan Macmillan Australia.

Mackey, Brendan G., Heather Keith, Sandra Berry L. Berry, and David B. Lindenmayer. 2008. *Green Carbon: The Role of Natural Forests in Carbon Storage.* Canberra: ANU E-Press.

Maddison, Sarah. 2019. *The Colonial Fantasy: Why White Australia Can't Solve Black Problems.* Sydney: Allen & Unwin.

Magdoff, Fred and John Bellamy Foster. 2011. *What Every Environmentalist Needs to Know About Capitalism*. New York: Monthly Review Press.

Magdoff, Fred and Chris Williams. 2017. *Creating an Ecological Society: Toward a Revolutionary Transformation*. New York: Monthly Review Press.

Malm, Andeas. 2020. *Corona, Climate, Chronic Emergency: War Communism in the Twenty-First Century*. London: Verso.

Maloney, Michelle. 2019. Rights of nature, earth democracy and the future of environmental governance. In *Rebalancing Rights: Communities, Corporations & Nature*. Tim Hollo, ed. Pp. 11–24. Canberra: The Green Institute. www.greeninstitute.org.au, accessed February 14, 2021.

Manne, Robert. 2011. How can climate change denialism by explained. *The Monthly*, December 8.

Manning, Paddy. 2019. *Inside the Greens: The Origins and Future of the Party, the People and the Politics*. Melbourne: Black, Inc.

———. 2020. *Body Count: How Climate Change Is Killing Us*. Sydney: Simon & Schuster.

Mares, Peter. 2018. *No Place Like Home: Repair Australia's Housing Crisis*. Melbourne: Text Publishing.

Mathews, Freya. 1991. *Ecological Self*. London: Routledge.

Matthews, Graham. 2010. Too many cars: Expand public transport. *Green Left*, July 1, p. 12.

McAuley, Ian. 2010. Taxing the miners' uncommonly large profit. *Dissent* 33 (Spring): 20–25.

McDonald, Jan. 2014. Hot in the city: planning for climate change impacts in urban Australia. In *Four Degrees of Global Warming: Australia in a Hot World*. Peter Christoff, ed. Pp. 172–189. London: Earthscan from Routledge.

McDonald, Matt. 2016. Bourdieu, environmental NGOs, and Australian climate politics. *Environmental Politics* 25: 1058–1078.

McIntosh, Alastair. 2020. *Riders of the Storm: The Climate Crisis and the Survival of Being*. Edinburgh: Birlinn.

McKibben, Bill. 1989. *The End of Nature*. New York: Random House.

———. 2019a. *Falter: Has the Human Game Begun to Play Itself Out?* Melbourne: Black, Inc.

———. 2019b. The climate movement: What's next? Opening reflections for a GTI forum. Great Transition Initiative: Toward a Transformative Vision and Future, June. https://greattransition.org, accessed October 24, 2019.

McKnight, David. 2018. *Populism Now! The Case for Progressive Populism*. Sydney: NewSouth.

McNab, Duncan. 2021. *The Ruby Princess*. Sydney: Palgrave Macmillan.

Meadows, Donella H. et. al. 1972. *The Limits to Growth: A Report for the Club of Rome's Project on the Predicament of Mankind*. London: Earth Island.

Meat & Livestock Australia. 2020. State of the industry report: The Australian red meat and livestock industry. http://www.mla.com.au, accessed March 16, 2021.

Megalongenis, George. 2016. Australia between recession and renewal. *Quarterly Essay*, No. 6: 1–67.

Mercer, Alexandra, Kim de Rijke, and Wolfram Dressler. 2014. Silences in the boom: Coal seam gas, neoliberalizing discourse, and the future of regional Australia. *Journal of Political Ecology* 21: 279–302.

Merchant, Carolyn. 1996. *Earthcare: Women and the Environment*. New York: Routledge.

Merchant, Carolyn. 2005. *Radical Ecology: The Search for a Liveable World*. New York: Routledge.

Metz, Bert. 2010. *Controlling Climate Change*. Cambridge, UK: Cambridge University Press.

Michaels, Patrick. 2004. *Meltdown: The Predictable Distortion of Global Warming by Scientists*. Washington, DC: Cato Institute.

Mikler, John. 2013. Climate innovation: Australian corporate perspectives on the role of government. *Australian Journal of Politics and History* 59: 414–428.

Milner, Andrew and J.R. Burgmann. 2021. Anthropocene fiction and world-systems analysis. *Journal of World-System Research* 26(2): 350–371.

Monash Sustainable Development Institute. 2020. Transforming Australia SDG progress report: 2020 update. www.sdgtransformingaustralia.com, accessed March 27, 2021.

Moore, Jason W., ed. 2016. *Anthropocene or Capitalocene? Nature, History, and the Crisis of Capitalism*. Oakland, CA: PM Press.

Morgan, Gareth and John McCrystal. 2009. *Poles Apart: Beyond the Shouting and Who's Right about Climate Change?* Melbourne: Scribe.

Murphy, John. 1993. *Harvest of Fear: A History of Australia's Vietnam War*. Sydney: Allen & Unwin.

Murphy, Lily. 2019. 'We don't want a world that's shit'. *Green Left*, February 12, p. 10.

Nash, Daphne, Paul Memmott, Joseph Reser, and Samid Suliman. (2018). *We're the same as the Inuit*: Exploring Australian Aboriginal perceptions of climate change in a multidisciplinary mixed methods study. *Energy Research & Social Science* 45: 107–119.

Nelson, Anitra. 2018. *Small Is Necessary: Shared Living on a Shared Planet*. London: Pluto Press.

Net Zero Momentum Tracker. 2020. Resources sector. December. Climate Works Australia. www.netzerotracker.org, accessed February 16, 2021.

Newell, Peter and Matthew Paterson. 2010. *Climate Capitalism: Global Warming and the Transformation of the Global Economy*. Cambridge, UK: Cambridge University Press.

Newman, Peter. 2009. Transport opportunities: Towards a resilient city. In *Opportunities Beyond Carbon: Looking for a Sustainable World*. John O'Brien, ed. Pp. 98–115. Melbourne: Melbourne University Press.

Nichols, Dick. 2010. Five questions about a carbon tax. *Green Left*, March 10, p. 11.

North, Peter. 2016. *Growing for Broke: How the Government Has Sold Out to Private Interests*. Berwick, VIC: Tomorrow Press.

Nursey-Bray, Mellissa, R. Palmer, T.F. Smith, and P. Rist. 2019. Old ways for new days: Australian Indigenous peoples and climate change. *Local Environment* 24: 473–486.

O'Connor, Mark and William J. Lines. 2008. *Overloading Australia: How Governments Dither and Deny on Population*. Sydney: Envirobook.

Oakley, Corey. 2013. What kind of organisation do socialists need? *Marxist Left Review* 5: 1–22.

———. 2019. Solid result for Victorian Socialists in first federal election campaign. *Green Left*, No. 1222, May 24.

OECD. 2017. Average annual hours worked per worker. www.stats.oecd.org, accessed January 21, 2021.

———. 2021. Trade unions. February 22. http://oecd.org, accessed March 10, 2021.

One Nation. 2019. Climate change. http://onenation.org.au/climate-change, accessed January 12, 2021.

Orgill, Brad. 2013. *Why Labor Should Savour Its Greens: Rebuilding a Fractured Alliance*. Melbourne: Scribe.

Palutikoff, Jean J., Jon Barnett, and Daniel A. Guttart. 2014. Can we successfully adapt to four degrees of warming? Yes, no and maybe. In *Four Degrees of Global Warming: Australia in a Hot World*. Peter Christoff, ed. Pp. 207–215. London: Earthscan from Routledge.

Pariona, Amber. 2017. Countries with the largest ecological footprints. *World Atlas*. htpp://www.worldatlas.com/articles/countries-with-the-largest-ecological footprint.html

Parliamentary Library. 2019. *Overseas Students in Australian Higher Education: A Quick Guide*. Canberra, ACT: Department of Parliamentary Services, Parliament of Australia.

Patel, Raj and Jason W. Moore. 2018. *History of the World in Seven Cheap Things: A Guide to Capitalism, Nature, and the Future of the Planet*. Melbourne: Black, Inc.

Paul, Erik. 2012. *Neoliberal Australia and US Imperialism in East Asia*. New York: Palgrave Macmillan.

———. 2016. *Australian Political Economy of Violence and Non-Violence*. London: Palgrave Macmillan.

Pearman, Graeme I., ed. 1980. *Carbon Dioxide and Climate – Australian Research*. Canberra: Australian Academy of Science.

Pearse, Guy. 2009. Coal, climate change and the end of the resources boom. *Quarterly Essay*, No. 33: 1–122.

Pearse, Guy, David McKnight, and Bob Burton. 2013. *Big Coal: Australia's Dirtiest Habit*. Sydney: NewSouth.

Percy, John. 2005. *Resistance: A History of the Democratic Socialist Party and Resistance, Volume 1: 1965–72*. Sydney: Resistance Books.

———. 2017. *Against the Stream: The Socialist Workers Party 1972–92*. Melbourne: Interventions.

Peterson, Dave. 2020. Top 10 richest person in Australia in 2020. *Unwrapped*. www.australiaunwrapped.com/top-10-richest-persons-in-australia-in-2020/

Pettifor, Ann. 2019. *The Case for the Green New Deal*. London: Verso.

Phibbs, Peter and Nicole Gurran. 2018. Housing in Australia: The game of homes. In *Wrong Way: How Privatisation & Economic Reform Backfired*. Damian Cahill and Phillip Toner, eds. Pp. 202–220. Melbourne: La Trobe University.

Pietsch, Sam. 2005. To have and to hold onto: Wealth, power and the capitalist class. In *Class and Struggle in Australia*. Rick Kuhn, ed. Pp. 21–38. Sydney: Pearson Longman.

Pitron, Guillaume. 2020. *The Rare Metals War: The Dark Side of Clean Energy and Digital Technologies*. Melbourne: Scribe.

Quantock, Rod. 1999. *Double Dissolution: Musings from a Rebel Comic*. Melbourne: Lothian Books.

———. 2008. *Rod Quantock's Guide to the Wonderful World of Climate Change*. Melbourne: Rod Quantock.

Quickie, Audrey. 2019. HeatWatch: Extreme heat in the Kimberley. Australia Institute, November.

Quickie, Audrey and Ebony Bennett. 2020. Climate of the nation 2020: Tracking Australia's attitudes towards climate change and energy. Australian Institute report.

Quiggan, John. 2017. The economic (non)viability of the Adani Galilee Basin Project. Australia Institute. htpp://www.tai.org.au/content/economic-nonviablity-adani-basin-project.

Radin, Paul. 1920. *The Autobiography of a Winnebago Indian*. Berkeley: University of California Press.

Randers, Jorgen and Paul Gilding. 2010. The one degree war plan. *Journal of Global Responsibility* 1(1): 170–188.

Reisinger, A., R.L. Kitching, F. Chiew, L. Hughes, P.C.D. Newton, S.S. Schuster, A. Tait, and P. Whetton. 2014. Australasia. Climate change 2014: Impacts, adaptation, and vulnerability, part B: Regional aspects. *Contribution of Working Group II to the Fifth Assessment Report of the International Governmental Panel on Climate Change*. V.R. Barrios, C.B. Field, D.J. Dokken, M.D. Mastrandrea, K.J. Mach, T.E. Biior, M. Chatterjee, K.I. Ebi, Y.O. Estrada, R.C. Genova, B. Girma, E.S. Kissell, A.N. Levy, S. MacCracken, P.R. Mastrandrea, and L.L. White, eds. Pp. 1371–1438. Cambridge, UK: Cambridge University Press.

Reucassel, Craig. 2020. *Fight for Planet A: The Climate Challenge and What We Can Do When There's No Planet B*. Sydney: ABC Books.

Rhiannon, Lee. 2012. What's the future for the Australian Greens? In *Left Turn: Political Essays for the New Left*. Antony Loewenstein and Jeff Sparrow, eds. Pp. 27–40. Melbourne: Melbourne University Press.

Rickards, Lauren and Karlie Tucker. 2009. Challenges for Australian agriculture. In *Climate Change for Young and Old*. Helen Sykes, ed. Pp. 84–101. Sydney: Future Leaders.

RMIT ABC Fact Check Unit. 2019. Do exports push Australia's greenhouse gas emissions to more than 5% of global totals. November 25. www.crikey.com.au/2019/11/25/fact-check-australias-greenhouse.

Robinson, Geoff. 2019. *Being Left-Wing in Australia: Identity, Culture and Politics after Socialism*. Melbourne: Australian Scholarly Publishing.

Rogelji, J. et al. 2016. Paris Agreement climate proposals need a boost to keep warming well below 2°C. *Nature* 534(7609): 631–639.

Rosewarne, Stuart, James Goodman, and Rebecca Pearse. 2014. *Climate Action Upsurge: The Ethnography of Climate Movement Politics*. London: Routledge.

Ross, Liz. 2011. Dealing with climate change. *Marxist Left Review*, No. 3: 101–120.

Ruhanen, Lisa and Aishath Shakeela. 2014. Responding to climate change: Australian tourism industry perspectives on current challenges and future directions. In *Climate Change and Tourism in Asia Pacific*. Bruce Prideaux, Bob McKercher, and Karen Elizabeth McNamara, eds. Pp. 35–51. London: Routledge.

Runyan, William McKinley. 1982. *Life Histories and Psychobiography: Exploration in Theory and Method*. New York: Oxford University Press.

Salamon, Margaret Klein with Molly Gage. 2020. *Facing the Climate Emergency: How to Transform Yourself with Climate Truth*. Gabriola Island, BC: New Society Publishers.

Salleh, Ariel. 1997. *Ecofeminism as Politics: Nature, Marx and the Postmodern*. London: Zed Books.

———. 2009. From eco-sufficiency to global justice. In *Eco-Sufficiency & Global Justice: Women Write Political Ecology*. Ariel Salleh, ed. Pp. 291–312. London: Pluto Press.

Sarkar, Saral. 1999. *Eco-Socialism or Eco-Capitalism? A Critical Analysis of Humanity's Fundamental Choices*. London: Zed Books.

Schweickart, David. 2016. Economic democracy: An ethically desirable socialism that is economically viable. The Next System Project. http://nextsystemproject.org

Seamer, Peter. 2019. *Breaking Point: The Future of Australian Cities*. Melbourne: Nero.

Seed. n.d. http://www.seedmob.org.au/join.

Servigne, Pablo, Raphael Stevens, and Gauthier Chapelle. 2021. *Another End of the World Is Possible*. London: Polity.

Shannon, Thomas R. 1996. *An Introduction to the World-System Perspective* (2nd Edition). Boulder, CO: Westview Press.

Shields, Robin. 2019. The sustainability of international higher education: Student mobility and global climate change. *Journal of Cleaner Production* 217: 594–602.

Shiva, Vandana. Maria Mies, and Ariel Salleh. 2014. *Ecofeminism*. London: Zed Books.

Silber, Irwin. 1994. *Socialism: What Went Wrong? An Inquiry into the Theoretical and Historical Sources of the Socialist Crisis*. London: Pluto Press.

Sklair, Leslie. 1991. *Sociology of the Global System*. Baltimore, MD: Johns Hopkins University Press.

———. 2002. *Globalization: Capitalism and Its Alternatives*. Oxford, UK: Oxford University Press.

Smith, Richard. 2020. *China's Engine of Environmental Collapse*. London: Pluto Press.

Socialist Alliance. 2020. Ecosocialist Manifesto: For an egalitarian, cooperative road to an ecosocialist future. *Green Left*, September 29, p. 11.

Sparrow, Jeff. 2005. The workers' flag is deepest green: Class struggles and the environment. In *Class and Struggle in Australia*. Rick Kuhn, ed. Pp. 196–212. Sydney: Pearson Education Australia.

Spratt, David and Philip Sutton. 2008a. *Climate Code Rd: The Case for a Sustainability Emergency*. Melbourne: Friends of the Earth Australia.

_____. 2008b. *Climate Code Red: The Case for Emergency Action*. Melbourne: Scribe.

Starr, Amory. 2005. *Global Revolt: A Guide to the Movements Against Globalization*. London: Zed Books.

Steffen, Will et al. 2009. *Australia's Biodiversity and Climate Change*. Melbourne: CSIRO Publishing.

Stilwell, Frank. 2000. *Changing Track: A New Political Economic Direction for Australia*. Sydney: Pluto Press Australia.

Stock, Petra, Will Steffen, Greg Bourne, and Louis Brailsford. 2018. *Waiting for the Green Light: Transport Solutions to Climate Change*. Canberra: Climate Council.

Swann, Tom. 2019. High carbon from a land down under: Quantifying CO_2 from Australia's fossil fuel mining and exports. The Australia Institute, July.

Talukdar, Ruchira. 2016. Hiding neoliberal coal behind the Indian poor. *Journal of Australian Political Economy*, No. 78: 132–158.

Tangney, Peter. 2019. Between conflation and denial – the politics of climate expertise in Australia. *Australian Journal of Political Science* 54(1): 131–149.

Taylor, Maria. 2014. *Global Warming and Climate Change: What Australia Knew and Buried . . . Then Framed a New Reality for the Public*. Canberra, ACT: ANU Press.

Thornett, Alan. 2019. *Facing the Apocalypse: Arguments for Ecosocialism*. London: Resistance Books.

Tietze, Tad and Elizabeth Humphreys. 2012. In *Left Turn: Political Essays for the New Left*. Antony Loewenstein and Jeff Sparrow, eds. Pp. 10–26. Melbourne: Melbourne University Press.

Toohey, Brian. 2019. *Secret: The Making of Australia's Security State*. Melbourne: Melbourne University Press.

Torstad, Vegard H. 2020. Participation, ambition and compliance: Can the Paris Agreement solve the effectiveness trilemma? *Environmental Politics* 29: 761–780.

Trainer, Ted. 1989. *Developed to Death: Rethinking Third World Development*. London: Green Print.

_____. 1991. *A green perspective on inequality*. In *Inequality in Australia: Slicing the Cake – the Social Justice Collective*. Jan O'Leary and Rachel Sharp, eds. Pp. 105–124. Melbourne: William Heinemann Australia.

_____. 1995. *The Conserver Society: Alternatives for Sustainability*. London: Zed Books.

_____. 1998. *Saving the Environment: What Will It Take?* Sydney: University of New South Wales Press.

_____. 2007. *Renewable Energy Cannot Sustain a Consumer Society*. New York: Springer.

_____. 2010. *The Transition to a Sustainable and Just World*. Sydney: Envirobook.

_____. 2019. The simpler way: Housing, living and settlements. In *Housing for Degrowth: Principles, Models, Challenges and Opportunities*. Anitra Nelson and Francois Schneider, eds. Pp. 120–130. London: Earthscan from Routledge.

Turner, George. 1987. *The Sea and the Summer*. London: Faber and Faber.

United Nations World Tourism Organization and International Transport Forum. 2019. *Transport-related CO_2 Emissions of the Tourism Sector: Modelling Results*. Madrid: World Tourism Organization. www.e-unwto.org/doi/book/10.18111/97892844166602021, accessed February 9, 2021.

Universities Australia. 2019. Higher education: Facts and figures. July. www.universities australia.edu.au, accessed April 15, 2020.

———. 2020. Data snapshot. htpp://www. universitiesaustralia.edu.au, accessed November 21, 2020.

U.S. Energy Information Administration. 2007. Australia energy statistics and analysis. January 24. http//www.eia.gov/international/overview/country/AUS.

Van Dijk, Albert I.J., et al. 2013. The Millennium drought in southeast Australia (2001–2009): Natural and human causes and implications for water resources, ecosystems, economy, and society. *Water Resources Research* 49: 1040–1057.

Victorian Socialists. 2018. *Manifesto*. Melbourne: Victorian Socialists.

Wall, Derek. 2010. *The No-Nonsense Guide to Green Politics*. Oxford: New Internationalist.

Wallace-Wells, David. 2019. *The Uninhabitable Earth: The Story of the Future*. London: Allen Lane.

Wallerstein, Immanuel. 2004. *World-Systems Analysis: An Introduction*. Durham, NC: Duke University Press.

Wangan and Jagalingou Family Council. n.d. http://www.wanganjagalingou.com.au/our-fight/, accessed July 17, 2019.

Warren, Matthew. 2019. *How Is Energy-Rich Australia Running out of Electricity*. Melbourne: Affirm Press.

Wettenhall, Roger. 1965. Public ownership in Australia. *Political Quarterly*, No. 36: 426–440.

Whetton, Penny, David Karoly, Ian Watterson, Leanne Webb, Frank Drost, Dewis Kirono and Kathleen McInnes. 2014. Australia's climate in a Four Degree World. In *Four Degrees of Global Warming: Australia in a Hot World*. Peter Christoff, ed. Pp. 17–32. New York: Routledge.

Whyte, David. 2020. *Ecocide: Kill the Corporation Before It Kills You*. Manchester: Manchester University Press.

Wijkman, A. and Johann Roeckstrom. 2011. *Bankrupting Nature: Denying the Planetary Boundaries*. London: Earthscan.

Wilkinson, Marian. 2020. *The Carbon Club: How a Network of Influential Climate Sceptics, Politicians and Business Leaders Fought to Control Australia's Climate Policy*. Sydney: Allen & Unwin.

World Coal Association. 2015. Coal marking and pricing. www.worldcoal.org/coal/coal-market-pricing, accessed November 22, 2015.

World Population Review. 2021. Greenhouse gas emissions by country 2021. http://world-populationreview.com, accessed March 17, 2021.

Wright, Christopher and Daniel Nyberg. 2015. *Climate Change, Capitalism, and Corporations: Processes of Creative Self-Destruction*. Cambridge, UK: Cambridge University Press.

Wright, Erik Olin. 2019. *How to Be an Anti-Capitalist in the 21st Century*. London: Verso.

Yoo, Tony. 2020. The highest paid CEOs in the ASX 200. *Motley Fool*, August 19. www.fool.com.au, accessed March 11, 2021.

Zizek, Slavoj. 2020. *A Left That Dares to Speak Its Name*. Cambridge, UK: Polity.

INDEX